The Inverse Hall-Petch Problem

David J. Fisher

Published by **Materials Research Forum LLC**
Millersville, PA 17551, USA

Published as part of the book series
Materials Research Foundations
Volume 55 (2019)
ISSN 2471-8890 (Print)
ISSN 2471-8904 (Online)

Print ISBN 978-1-64490-034-5
ePDF ISBN 978-1-64490-035-2

Distributed worldwide by

Materials Research Forum LLC
105 Springdale Lane
Millersville, PA 17551
USA
http://www.mrforum.com

Printed in the United States of America
10 9 8 7 6 5 4 3 2 1

Table of Contents

Introduction

It was tempting to add a triumphant 'Solved' to the title of the present work because recent research results have suggested that the problem in question has a solution. But first the problem has to be laid out.

The Hall-Petch equation is one of the most familiar and useful empirical equations of materials science, with citations of the original papers running into many thousands. The strengthening effect which it describes is also one of the main tools at the disposal of the metallurgist for improving the mechanical properties of a polycrystalline material. It is based upon the seminal works[1,2] of Eric Ogilvie Hall and Norman James Petch and is of the form,

$$\sigma_y = \sigma_0 + kd^{-1/2}$$

where σ_0 and k are constants and d is the grain size of a material. It is quite instructive to read the original papers, especially that of Hall, because the work was done when the dislocation concept was still in its infancy and the theory of grain-boundary effects was still very much in the thrall of older ideas, and there was much talk of grain-boundary 'films'. The dislocation pile-up model that was pioneered in papers such as Hall's has endured ever since. In the conclusion to his paper, Hall mused on the possible effect of the segregation of alloying elements to the boundaries; a topic which will be found to be very relevant to the present state of understanding of the phenomenon. An underlying problem is that the hardening might be caused not only by a reduction in the grain-size but also by relaxation of the non-equilibrium grain-boundary structure. The latter is difficult to evaluate quantitatively but the recent apparent solution to the inverse-Hall-Petch problem probably owes a great deal to a new understanding of, and ability to control, grain boundary relaxation.

The Normal Hall-Petch Effect

Originally formulated for mild steel, the Hall–Petch equation has been shown to apply to pure metals, intermetallics and multi-phase alloys. Because a material *obviously* becomes stronger as the grain size decreases, according to the above equation, it was only natural to assume that nanocrystalline materials, when they first became a new feature of materials science[3], would automatically possess high strengths thanks to their minute grain sizes. That small size unfortunately also makes them vulnerable to other phenomena; ones which tended to weaken them when the grain size is too small. The aim of the present work is to show how all of the previous research on the Hall-Petch Effect has very recently led to the achievement of the hoped-for levels of strength of nanocrystalline materials.

Materials Research Forum LLC
https://doi.org/10.21741/9781644900352

Just to play devil's advocate, it should be pointed out that the experimental evidence for the normal Hall-Petch law has received some criticism[4] in recent years with regard to the original data and, implicitly, to deviations from the normal law. It has been noted that, although data are traditionally fitted using a reciprocal square-root function they might equally well be fitted using other functions; power-law and otherwise. Supposed long-term problems with the reciprocal square-root expression have been attributed to artefacts of defective data analysis. Bayesian meta-analysis, itself a slightly disreputable technique, supposedly shows that the data can be closely fitted by using a simple reciprocal, or other function. These alternative functions are also supposedly supported by theory. It has been concluded that the Hall-Petch effect is not explained by the various extant theories, but is a manifestation of a general scaling relationship between the required stress and the available space in which dislocation sources must operate. A further study of published values of the Hall-Petch coefficients from 61 sets of data revealed[5] no dependence of the coefficients upon plastic strain, but a strong dependence upon the average grain-size in each set. Some 80% of the 61 coefficients could be accounted for. It was again concluded that the grain-size dependence of the Hall-Petch coefficient is an artefact arising from fitting data to an inappropriate function. An alternative grain-size effect involving a $\ln[d]/d$ function was considered to be more consistent with theories of dislocation dynamics and generation.

Because the original work of Hall and Petch was related to iron-based alloys, it is logical to start there in order to see how the then-new phenomenon affected such materials.

Iron

In an early study[6] of nanocrystalline materials, iron powder of 3N-purity was milled for up to 220h; the hardness increasing continuously with milling time and refinement of the cell/grain structure. The flow-stress results, assuming a Tabor ratio of 3.3, could be fitted by a Hall-Petch equation of the form,

$$\sigma_f(MPa) = 371 + 685d(\mu m)^{-1/2}$$

It was noted however that the data could also have been fitted by using a d^{-1} relationship with just about the same degree of confidence. It was noted in an early review[7] that, during mechanical milling, the treated powder stored up an extremely high strain-energy, resulting in a marked work-hardening as well as forming nanocrystalline grains. In the case of iron the hardness, up to that date, had attained 1024(DPH) and the grain-size had been reduced to 3.4nm. Polycrystallization of dislocation cells and sub-grains also occurred during grain-refining. When the milled powders were annealed, they underwent various microstructural changes, depending upon their degree of work-hardening. When moderately work-hardened, normal recovery and recrystallization occurred with increasing annealing temperature. When they were work-hardened to an extent corresponding to the smallest attainable grain-size, normal grain-growth alone occurred during annealing and led to softening of the powder. Under average milling conditions, the degree of work-hardening was of an intermediate nature and both of the above extremes were possible, leading to some overlap. It was noted that, for both milled and annealed iron powders, the relationship between hardness and grain-size closely obeyed the Hall-Petch law for a wide range of grain-sizes, down to 6nm. The microstructure of mechanically processed powders also depended upon the process details. Lengthy high-energy attrition ball-milling in argon produced nanocrystalline body-centred cubic grains, while milling in nitrogen or alloying with carbon produced micron-sized particles with nanosized body-centred cubic grains plus smaller nanosized body-centred tetragonal grains, with the carbon and nitrogen concentrated at the grain boundaries. The addition of aluminium had little effect upon the microstructure[8]. Explosive consolidation produced dense compacts having a microstructure which was similar to that of mechanically processed powder. The hardness of compacts as a function of grain-size obeyed the Hall-Petch law, but it was concluded[9] that the processing conditions had only a secondary effect upon the hardness, and more strongly affected the thermal stability of the microstructure. Nanocrystalline thin films, prepared[10] by sputtering, had grain-sizes of 10.3, 10.5, 11.4, 11.7, 12.4, 16.7, 18.9 or 21.0nm. The grain-size was varied by altering the sputter-gun to substrate distance, and it was not possible to grow relatively pure as-deposited films having smaller or larger grain-sizes. The hardness of the as-deposited films could be described by,

$$H(GPa) = 2.26 + 0.152d(\mu m)^{-1/2}$$

Because of a large scatter and rather poor fit however, little attention was paid to these results. The effect of grain-size upon the ductility of nanocrystalline iron which had been ball-milled and consolidated using uniaxial warm-pressing to give a range of grain-sizes was studied[11]. The hardness as a function of grain-size was described by a Hall-Petch slope which was smaller than that for coarse-grained iron. In tension, the failure was macroscopically brittle, with local ductility observed within very concentrated shear bands. The compressive behaviour of nanocrystalline samples was similar to that of perfectly plastic material. When nanocrystalline powders with a grain-size of 13nm were prepared[12] by ball-milling, and consolidated by spark plasma sintering (550C, 80 to 220MPa), the average grain-size of the consolidated material was about 80nm and was independent of the sintering time at constant pressure. The hardness of super fine-grained material, sintered for 600s under 220MPa, was 470H$_v$. This value fell on the extrapolated Hall-Petch plot of the above Malow-Koch data for grain-sizes of less than 40nm. *In situ* transmission electron microscopy was used[13] to investigate the deformation mechanisms operating in physical vapour-deposited nanocrystalline body-centred cubic iron films with an average grain-size of 35nm. Tensile straining showed that fracture occurred at strains of about 5% and that cracks propagated mainly via the separation of grain boundaries at the crack tip, together with localized intragranular ductile fracture. The crack-tip deformation in turn involved dislocation-motion, grain-rotation and grain-growth. But there was no sign of twinning. The concurrence of grain-rotation and dislocation-motion indicated that grain-rotation occurred at fairly large grain sizes, and there was no sharp transition between dislocation-mediated and grain-boundary sliding mechanisms as the grain-size decreased.

The effect of the ferrite grain size upon dislocation-strengthening was investigated[14] in carbon-steels containing 0.006 to 0.15%C, and having grain-sizes ranging from 1 to 100μm. In slightly deformed specimens, the dislocation density increased in proportion to the reciprocal of the ferrite grain-size. At dislocation densities of below $2 \times 10^{14}/m^2$, the density increased linearly with deformation strain, but started to level-off due to dynamic dislocation-recovery when the density exceeded that level. Tensile tests revealed meanwhile that the yield stress obeyed the Hall-Petch law in the case of as-annealed material, but obeyed the Bailey-Hirsch relationship in the case of cold-rolled specimens. This implied that the flow-stress depended only upon the dislocation density; regardless of the grain-size. It was concluded that the introduction of dislocations had been promoted by the decreased ferrite grain-size and that this then resulted in an increase in flow-stress in the uniform-deformation range.

The so-called quenching-and-partitioning steels combine excellent strength and ductility, but their microstructures can contain carbon-depleted martensite, (initial martensite), bainitic ferrite, secondary martensite and retained austenite. Modeling the yield strength of such microstructures is therefore difficult without knowing the yield strength of each phase. The microstructures have most recently[15] been simplified, and treated as a mixture of α'-phase (carbon-depleted martensite, secondary martensite or bainitic ferrite) and γ-phase (retained austenite). Two generalized physical models could then predict the yield strength of as-quenched and quenching-and-partitioning steels with carbon contents ranging from 0.06 to 0.42wt%. These models showed that the yield strength of the α'-phase varied as the reciprocal of the square-root of the block width (effectively a Hall-Petch relationship), but varied as the reciprocal of the lath thickness (the Langford-Cohen relationship). The assumptions and simplifications made with regard to the microstructure did not greatly affect the accuracy of the predicted yield strengths of either as-quenched steel or quenching-and-partitioning steels; the error being less than 10%. It was concluded that residual carbon in the α'-phase was still the main factor which

Materials Research Forum LLC
https://doi.org/10.21741/9781644900352

controlled the yield strength of quenching-and-partitioning steels, in spite of the possibly very uneven carbon distributions.

Specimens of carbon-supersaturated nanocrystalline hypereutectoid steels having a tensile strength of 6.35GPa were produced[16] from severely cold-drawn pearlite. The material softened during annealing at between 200 and 450C, and the elongation-to-failure exhibited a non-monotonic dependence upon temperature. Sub-grain coarsening occurred during annealing, and led to a strength reduction in accord with the Hall-Petch law. Analysis of the manganese distribution near to sub-grain boundaries, and in the cementite, revealed signs of capillarity-driven sub-grain coarsening via sub-grain boundary migration. A marked deterioration in ductility after annealing at above 350C was attributed to the formation of cementite at sub-grain boundaries.

Ultra fine-grained specimens of Fe-Cr-Mo-V AISI H13 tool steel were produced[17] by high-energy mechanical milling and spark plasma sintering. Bimodal grain-size microstructures were created by mixing various fractions of nanocrystalline and coarse-grained powder. Dislocation strengthening (435MPa) and Hall-Petch strengthening (388MPa) were the main contributors to the overall strength of the as-sintered steel. The presence of 34vol% of retained austenite had a negative effect upon the strength (555MPa). The recovery of dislocations and defects which was associated with the precipitation of carbides reduced the effect of grain-size upon the strength after tempering.

Nanocrystalline and ultra fine-grained microstructures were obtained[18] in Fe-17Cr-6Ni austenitic steel by combining severe cold-deformation with reverse-transformation annealing. Samples having a minimum average grain-size of about 220nm were obtained which exhibited a high strength and a high ductility following 75% cold-reduction and annealing at 700C for 20s. The relationship between grain-size and yield strength closely obeyed the Hall-Petch relationship down to a grain-size of 200nm. In the case of coarse-grained steel, the strain-hardening rate behaviour involved 4 stages and an increase in strain-hardening rate was attributed to deformation-induced martensite transformation. In the case of nanocrystalline and ultra fine-grained austenitic steel, the strain-hardening rate behaviour involved only 3 stages and the increase in strain-hardening rate was attributed to the effect of deformation-induced martensite transformation and twinning.

Specimens of 316L were produced[19] by high-pressure torsion, with the initial material being essentially single-phase γ-austenite with a coarse-grained (42μm) microstructure. The grain-size was reduced to about 45nm after 10 torsional turns, and a phase transformation transformed the initial γ-austenite, first to ε-martensite and finally to α'-martensite, with increasing strain. The dislocation density was of the order of $10^{16}/m^2$ in the main α'-martensite phase following the 10 turns. Formation of the multiphase nanocrystalline microstructure produced a four-fold increase in hardness, to a value of about 6000MPa. The Hall-Petch behaviour of the torsion-processed alloy was compared with that of coarse-grained samples (figure 1). In each case, the hardness versus grain-size results obeyed the Hall-Petch law but the slopes were different for the coarse-grained and ultra fine-grained ranges. It was concluded that the hardness was more sensitive to the grain-size than to the phase composition.

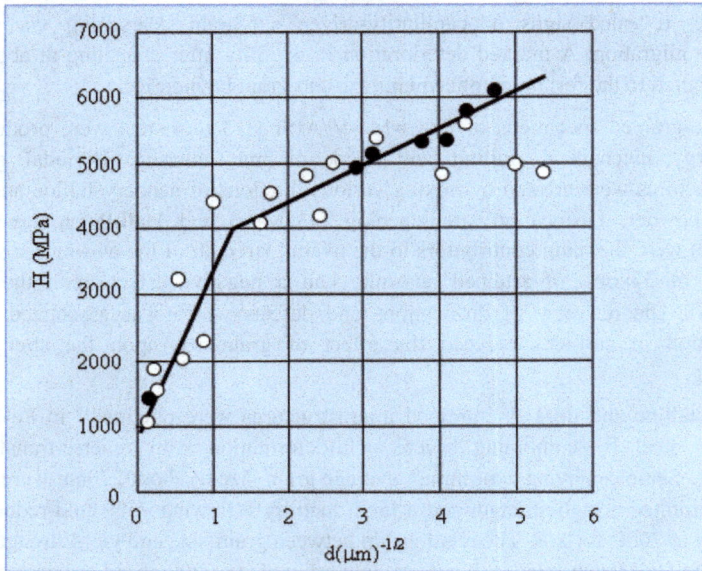

Figure 1. Hardness of AISI316L as a function of grain-size
Filled circles: present work, open circles: previous data

The thermal stability of the microstructure and the phase composition were determined[20] for the stainless steel after a ¼-turn or after 10 turns or torsional deformation. Differential scanning calorimetry thermograms contained two characteristic peaks. One peak was exothermic and was attributed to recovery of the dislocation structure, with no change in phase composition or grain-size. The activation energies required for recovery following a ¼-turn and 10 turns were about 163 and 106kJ/mol, respectively. This suggested that the process was controlled by diffusion along grain boundaries and dislocations. The second peak was endothermic and was attributed to a reverse transformation of α'-martensite to γ-austenite. The hardness of the annealed samples was due mainly to the grain-size and obeyed a Hall–Petch law of the form,

$$H(MPa) = 3220 + 580d(\mu m)^{-1/2}$$

The nanocrystalline material exhibited good thermal stability, with a grain-size of about 200nm after annealing (1000K) and a hardness of about 4900MPa. A detailed study was made[21] of the Hall-Petch behaviour of specimens of this stainless steel which had a yield strength of about 2.0GPa. A useful comparison was made of the Hall-Petch laws for this steel in various grain-size regimes (figure 2).

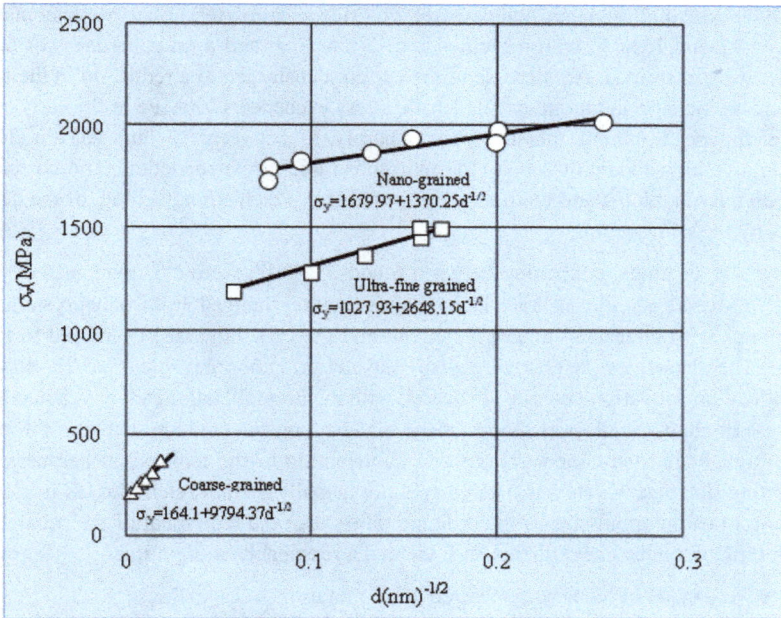

Figure 2. Yield strength of AISI316L as a function of grain-size

Supersonic fine-particle bombardment has been used[22] to create a 25μm-thick nanocrystalline surface layer on quenched and tempered 38CrSi steel. The grain-size of the upper surface layer was about 16nm, and increased with increasing depth from the surface. Nano-indentation testing showed that the mechanical properties of the surface layer were improved by the grain refinement and that its hardness was about twice that of the matrix. The hardness increased in accord with the Hall-Petch law, and the improvement in the mechanical properties of the treated layer was attributed mainly to the grain refinement. The elastic modulus of the surface layer was slightly affected by the treatment. Low-temperature annealing rendered uniform and stable the structure and properties of the nanostructured surface layer.

The relationship between microhardness and grain-size of nanocrystalline $(Fe_{0.99}Cu_{0.01})_{78}Si_9B_{13}$ was proved[23] to obey the Hall-Petch law for grain-sizes of 30 to 90nm[24].

Deposits of nanocrystalline Fe-Ni which were up to 400μm thick were laid down under high over-potential from an additive-free chloride electrolyte[25]. Specimens containing between 12 and 15%Ni, with a grain-size of 5.5 to 7nm and a microhardness of 647 to 711H_V, were examined. An increase in the current density led to a reduction in the grain-size and an increase in hardness. The tensile stress exceeded 1400MPa in the early stages of electrodeposition, but this decreased sharply to between 70 and 90MPa due to cracking. A higher current-density caused more rapid stress-reduction. The correlation between microhardness and grain-size confirmed Hall-Petch strengthening, of the deposit matrix by ultra-fine grains.

When Fe-W coatings, containing between 6 and 25at%W, were electrodeposited[26] from an Fe(III)-based glycolate-citrate bath, their structure changed from nanocrystalline to amorphous with increasing tungsten content, and the crystallite size decreased to below 10nm. The deposition temperature also played an important role, due to differing crystallization activation energies. In accord with the normal Hall–Petch law, a maximum hardness of about 10GPa was found for the highest tungsten content, with Fe_2W forming at the higher tungsten concentrations and contributing to the increase in hardness. The supporting diagram which was offered was not actually a Hall-Petch plot, as it used the square-root of the grain-size as the ordinate rather than the reciprocal of the square-root. Re-plotting of the data nevertheless still yielded a reasonably straight line.

Studies were made of other body-centred cubic metals

Chromium

Nanocrystalline chromium was produced[27] by electrodeposition, and specimens with grain-sizes of 19 to 57nm resulted from annealing the as-deposited material. The strength obeyed the normal Hall-Petch relationship, and it was suggested that the hardening due to grain refinement is generally greater for body-centered cubic metals than for face-centered cubic or hexagonal close-packed metals.

Tantalum

It was noted[28] that a ten-fold increase in the hardness of nanocrystalline tantalum and vanadium nanolaminates could be attributed to grain-size effects rather than to the layer-pair composite spacing. The hardness of the body-centred cubic nanocrystalline structures was clearly described by a Hall-Petch relationship. The spall strength in the dynamic failure of tantalum during laser-shock compression exhibits an appreciable dependence upon the grain size[29]. This dependence was opposite to the normal Hall-Petch law because the spalling was largely intergranular in both coarse-grained and nanocrystalline samples. That is, monocrystals exhibited a higher spall-strength than did polycrystals and these were stronger than were ultra fine-grain material. Ductile failure occurred via the nucleation, growth and coalescence of voids. In the case of monocrystals, the voids grew in the interior, whereas nucleation was intragranular and intergranular in the case of coarse and ultra fine-grained crystals. The length of the geometrically-necessary dislocation which was required to form an intergranular void was lower than that required to form an intragranular void, for a given maximum diameter, and the required energy was consequently lower. There was also an increase in spall strength as the strain-rate was increased from 6×10^6 to 5×10^7/s.

Vanadium

The intriguing high-entropy alloys already exhibit a high hardness and strength in the coarse-grained form, so it is to be hoped that the Hall-Petch effect can lead only to further improvement. Single-phase nanocrystalline powders of the high-entropy alloy, VNbMoTaW, were prepared[30] by mechanical alloying. The resultant nanocrystalline materials had a body-centered cubic structure and an average grain size of about 6nm. High pressures and high temperatures were used to consolidate the nanocrystalline powders into bulk nanocrystalline high-entropy alloy samples. Bulk nanocrystalline alloy which was consolidated by using a temperature of 1150C had an average grain-size of about 30nm and a hardness of 11.4GPa. This was twice the hardness of that of the coarse-grained material. This ultra-high hardness was attributed to solid-solution hardening, grain-boundary hardening and dislocation-hardening.

Niobium

Amorphization due to mechanical attrition was observed upon milling Nb+Sn, Nb+NbSn$_2$ or Nb$_3$Sn crystalline powders. Mechanical alloying first occurred in Nb+Sn or Nb+NbSn$_2$ powder mixtures, leading to the synthesis of A15-structured Nb$_3$Sn. During continued milling, this transformed into an amorphous structure. Complete disordering of the ordered Nb$_3$Sn phase occurred after milling for about 1h, while amorphization first began after 5 to 6h. Energy differences between the crystalline and amorphous forms suggested that the driving force for crystalline-amorphous transformation arose from the sum of the disordering-energy and the grain-boundary energy of the nanocrystalline structure which developed during milling. The hardness of the hard brittle Nb_3Sn increased by about 20% before the onset of amorphization. This increase was attributed to the refinement of the grain structure, and it could be fitted by using the Hall-Petch law[31].

In the case of hexagonal close-packed materials

Cobalt

Electrodeposited nanostructured Co-Ni coatings having grain-sizes ranging from 11 to 23nm were examined[32]. The finest grain-size was obtained by deposition from a nickel sulphate bath. The coatings had a mixed, face-centred cubic and hexagonal close-packed, structure plus various morphologies and porosities. When porosity was accounted for, the calculated pore-free hardness obeyed the normal Hall-Petch law; even when the grain size was reduced to 11nm. Nanocrystalline $Co_{78}Ni_{22}$, having grain-sizes ranging from 5 to 35nm, was prepared[33] by high-speed jet-electrodeposition and annealing. The microhardness increased in agreement with the normal Hall-Petch law when the grain-size was within the above range. On the other hand, the coercivity increased with increasing grain-size for sizes ranging from 5 to 15.9nm, and decreased again for grain sizes of above 16.6nm.

Titanium

Specimens having nano- and sub-micron grains, and consisting of γ-TiAl and ξ-$Ti_5(Si,Al)_3$, were produced[34] by high-energy milling and hot isostatic pressing. The yield strength and hardness both exhibited a Hall-Petch dependence upon the grain size, leading to high flow and fracture stresses under compressions of up to 3GPa. With decreasing grain-size, the strain-hardening coefficient and compressive fracture strain decreased, and reached zero at a grain-size of about 150nm. Room-temperature deformation was mediated by dislocation-glide and mechanical twinning; the latter increasing in importance as the grain-size was further reduced. The influence of diffusion-controlled deformation was ruled out, even for intermetallics with a crystallite size as small as 50nm. Reactive and non-reactive sintering were compared[35] in the production of titanium aluminides by the mechanical alloying and hot-pressing of powders having a composition of Ti-50at%Al. A solid-solution of Ti(Al) was formed in the early stage of milling, and transformed to an amorphous phase during prolonged milling. This amorphous structure eventually transformed into a supersaturated hexagonal close-packed Ti(Al) solid-solution containing trace amounts of TiAl after 80h of milling. This then completely transformed to TiAl, Ti_3Al and $TiAl_3$ during additional milling for up to 100h, with particles of about 200nm present. Hot-pressed pre-alloyed powder exhibited a better density, hardness, homogeneity, yield stress and ductility than did reactively-sintered material. The main contribution to the yield stress of sintered pre-alloyed powder arose from nanometre-sized intermetallic grains, due to the Hall-Petch effect.

The microhardness of commercial-grade titanium VT1-0 samples having grain-sizes ranging from 35nm to 10μm, when measured between 77 and 300K, showed that the nanocrystalline samples which were produced by low-temperature rolling were stable with respect to thermal and mechanical effects[36,37]. The results were well-described by the Hall-Petch law, although the parameters depended upon the temperature. The latter fact suggested that the microplastic deformation was thermally-activated and dislocation-related, regardless of the grain-size. In similar work[38], multi-pass rolling of VT1-0 at near to liquid-nitrogen temperatures again led to grain-sizes ranging from 10μm to 35nm; associated with an almost two-fold increase in microhardness. As in the earlier work above, the grain-size dependence of the microhardness in a Hall-Petch plot consisted of two parts. The marked temperature-dependence of the microhardness suggested that the plastic deformation was thermally activated. The anisotropy of the yield strength of nanocrystalline VT1-0 with a grain-size of about 45nm was determined[39,40] by performing uniaxial compression at a plastic deformation rate of 3.5×10^{-4}/s and at temperatures of 4.2 to 300K with the compression-axis parallel or perpendicular to the cryorolling direction. The crystallites were found to be morphologically anisotropic. Calculated and experimental values of the yield strengths, of samples deformed parallel and perpendicular to the cryorolling direction, obeyed the Hall-Petch relationship as a function of crystallite size. The anisotropy of the yield strength was related to the anisotropic shape of the grains and crystallites. When commercial-purity material was subjected[41] to surface mechanical attrition, it was found that twinning occurred only in the early stages of deformation. The absence of twin–twin intersections thus suggested that twinning was not directly responsible for reducing the grain-size. Dislocation slip was instead the main deformation mode, and led to refinement of the microstructure via the formation of low-angle lamellar boundaries. Continuous dynamic recrystallization was the nanocrystallization mechanism. A combination of Hall–Petch and Taylor relationships could explain the strength, but only if separate Hall–Petch and Taylor parameters were applied to the nanocrystalline surface and to the severely deformed sub-surface. When large numbers of stacking-faults were created as a result of partial dislocations emitted from grain boundaries, the faults markedly strengthened the nanocrystalline films via dislocation-fault interactions[42]. This was associated with a Hall-Petch slope of about 20GPa√nm, and a strength of about 4.4GPa; almost 50% of the ideal strength. When pure titanium films were prepared by magnetron sputtering at a bias voltage of 0 to 140V, the microstructure was a composite structure consisting of an amorphous-like matrix with embodied nanocrystallites[43]. Crystallization improved with increasing bias voltage. The hardness exhibited a linear dependence upon grain-size, between 6 and 15nm, but with a relatively low slope.

When Ti-5Ta-1.8Nb alloy was subjected[44] to cryorolling, the grain-size decreased with increasing strain, from about 5μm in annealed samples to some 20nm at a strain of 2.3. The dislocation density first increased sharply to a strain of 0.69 and then decreased, with increasing strains of up to 2.3. The decrease in dislocation-density was attributed to the formation and interaction of pile-ups and the consequent annihilation of dislocations. Hall-Petch plots exhibited two different slopes: one in which there was a rapid increase in stress, and one in which there was a negligible increase in stress, with decreasing grain-size. This was again attributed to changes in dislocation-density and grain-size. Dimples and micro-voids were seen on the fracture surfaces of coarse-grained specimens, while nanocrystalline specimens had a mixture of shear bands and ductile dimples.

Molecular dynamics simulations were used[45] to model the microscopic deformation mechanism of polycrystalline TiAl as a function of grain-size and temperature. At grain-sizes of less than 8nm, the yield stress of the nanocrystalline alloy increased with increasing grain-size in accord with the inverse-Hall-Petch relationship. This was due mainly to the migration of grain boundaries and to grain-rotation. When the grain-size was greater than 8nm, the degree of sensitivity of the yield stress to the grain-size decreased, and dislocation-slip plus deformation twins in the grain interiors then dominated plastic deformation. The modulus also increased with increasing grain-size. The temperature affected the Young's modulus, in that increasing the temperature then increased the distance between atoms, thus decreasing the binding force between atoms and decreasing the modulus. With increasing temperature, the dislocation density also decreased and dislocation-emission at grain boundaries was delayed.

Zirconium

Microstructural changes and nano-indentation hardness variations in nanocrystalline zirconium as a result of rolling at cryogenic temperatures were investigated[46]. The total deformed-layer depth was greater than 600μm, and the average grain-size ranged from about 8nm in the uppermost surface layer to micron-sized in the coarse-grained matrix. This corresponded to a hardness-gradient of about 6.0 to 2.86GPa. Deformation bands formed in the early stages of deformation and, with increasing strain, dislocation cells formed within the deformation bands and finally transformed into nanograins. The Hall-Petch plot of hardness versus grain-size was non-linear, due to a change in the deformation mechanism from dislocation pile-up to grain-boundary sliding as the grain-size decreased.

Zinc

Nanocrystalline zinc coatings having average grain-sizes ranging from 6 to 32nm, corresponding to current-densities of 0.3 to 2.4A/cm^2, were prepared[47] on copper substrates by means of pulsed electrodeposition. With decreasing grain-size, the Vickers microhardness increased from below 0.5GPa to over 2.0GPa. The friction coefficient, against a Si_3N_4 ball in ambient air, meanwhile decreased from 0.18 to 0.05. The microhardness obeyed the normal Hall-Petch relationship.

Coatings of nanocrystalline Zn-Ni alloy were deposited from an alkaline glycinate bath which contained a saccharin additive that decreased the crystallite size from 68 to 13nm with increasing concentration. The coatings also had a colony-like morphology, and the colony-size increased with increasing current-density. Here however was an example of a system in which the hardness did not obey the Hall-Petch law at all[48]. Nanocrystalline Zn-Ni coatings with a grain-size of about 26nm were electrodeposited[49] onto carbon-steel substrates from an alkaline bath containing 5,5'-dimethylhydantoin as a complexing agent. Optimum coatings, containing 13 to 16wt%Ni, were obtained using a $Ni^{2+}/(Zn^{2+}+Ni^{2+})$ ratio of 0.32, a current-density of 2A/dm^2, a temperature of 50C and an agitation-speed of 1000rpm. The phase-structure, grain-size, microhardness and corrosion resistance depended directly upon the nickel content of the deposit. The phase-structure changed from a mixture of η-phase and γ-phase to a single γ-phase with (411) orientation. The grain-size decreased with increasing zinc content in the deposit. An increase in current-density could also decrease the grain size. But here again an increase in the microhardness owed much more to an increase in nickel content than to any Hall-Petch effect.

Face-centred cubic metals and alloys are however the most widely studied group with regard to the Hall-Petch law.

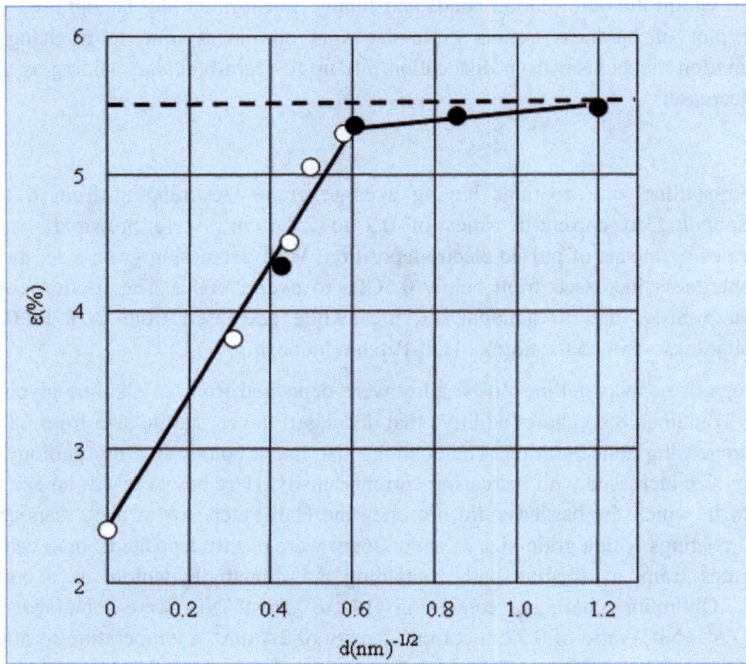

Figure 3. Hall-Petch behavior of the elastic strain limit of nano-twinned [111] gold nanowires. Open circles: experimental data, filled circles: molecular dynamics simulation. The dotted line indicates the theoretical elastic limit.

Gold

When nanocrystalline specimens having the theoretical density were prepared by gas deposition, the Young's modulus below 80K of specimens having a mean grain-size of 26 to 60nm was greater than 92% of the bulk value[50]. This suggested that the modulus in the grain-boundary region was greater than 70% of the bulk value. At above 200K, the anelastic strain due to relaxation in the grain-boundary region increased with increasing temperature, leading to a decrease in the Young's modulus. Plasticity tests which were performed at near to room temperature, using specimens having a mean grain-size of 15 to 60nm, revealed that the Hall-Petch relationship was obeyed by Vickers microhardness data with a holding-time of 10s, and that creep deformation occurred during quasi-static

Materials Research Forum LLC
https://doi.org/10.21741/9781644900352

tensile testing, indicating that plastic deformation at near to room temperature was governed by the stress level and by the strain-rate[51]. The response of free-standing nanocrystalline (40 to 100nm) films (0.25 to 1.00μm thick) was determined[52] by using a micro-tensile technique at strain-rates of 10^{-4} to $10^{-6}/s$[53]. The plastic properties were especially sensitive to the strain-rate, film thickness and grain-size although the elastic properties were relatively unaffected. Thinner films exhibited an appreciable strain-rate sensitivity while thicker films were only slightly affected. Hall-Petch boundary hardening occurred, plus controlled plastic flow at higher strain-rates, while diffusion-controlled deformation appeared to occur to an increasing degree as the strain-rate decreased. It was suggested that the films underwent predominantly power-law creep at lower strain rates for this grain-size. The introduction of nanoscale twins is effective in increasing the tensile strength of metals, but non-uniform twin sizes locally affect the yielding behavior. *In situ* high-resolution transmission electron microscopic tensile testing[54] of nanotwinned [111]-oriented gold nanowires revealed (figure 3) a clear Hall-Petch relationship in the elastic strain limit at up to 5.3%, near to the ideal theoretical limit, as the twin size decreased below 3nm. The largest twin, in nanowires containing irregular twin-sizes, controlled slip nucleation and yielding in pure tension. Continuous hardening, with no loss of strength, occurred in nano-twinned monocrystalline nanowires. This differed from the behavior of both bulk nanocrystalline and nano-twinned-nanocrystalline metals.

The study of electrodeposits showed that various samples had differing Vickers hardness values[55]. Upon comparing the grain size, the texture coefficient of the same crystal face and the Vickers hardness, it was found that abnormal hardness occurred when the grain-size was in the nanocrystalline range and the Hall-Petch relationship was here unsuitable for correlating the grain size with the hardness. Plated gold had a greater Vickers hardness when it had a stronger (111) orientation. The highest hardness values attained 117$H_{V0.2}$, and the maximum texture coefficient approached 70.59%. The orientation of the plating could be controlled by the use of additives during electrodeposition.

Nanocrystalline samples containing copper and/or tin were prepared[56] which had grain-sizes of down to about 3nm. Tensile measurements showed that the Hall-Petch law applied here right down to the smallest length-scale. It was noted that, although copper and tin solid solutions had little effect upon the strength, they markedly improved the thermal stability and abnormal grain-growth was no longer a characteristic feature of alloyed nanocrystalline gold. Gold-copper nanocrystalline alloys were electrodeposited[57] onto gold substrates from a gold-copper alkaline cyanide bath using direct current and temperatures of between 20 and 80C. Smooth, columnar or nodular coatings were obtained as a function of the copper content and deposit-thickness. The average grain-size of the alloys ranged from 3 to 35nm, and decreased with an increasing amount of copper in the coating. Alloy samples containing 12wt% of copper, with a grain-size of 6nm, exhibited a maximum microhardness of 378H$_V$. Specimens having grain sizes ranging from 6 to 35nm obeyed a direct Hall-Petch relationship.

Silver

The microhardness of nanocrystalline samples which were prepared by inert-gas condensation and *in situ* compaction was such that the values for specimens compacted using pressures ranging from 0.15 to 1.80GPa were greater than that for polycrystalline silver[58]. No effect, upon the microhardness, of internal strains produced by the compaction was detected. The microhardness nevertheless increased with increasing compaction-pressure and density. The effect of the grain size upon the microhardness was well-described by a normal Hall-Petch relationship for grain sizes ranging from 17 to 105nm (figure 4), with the linear normal portion being described by,

$$H_v(GPa) = 0.33 + 4.21d(nm)^{-1/2}$$

In the case where the microhardness increased with increasing grain size, the deviation from the Hall-Petch relationship was attributed to densification of the nanocrystalline silver rather than to an inverse Hall-Petch effect.

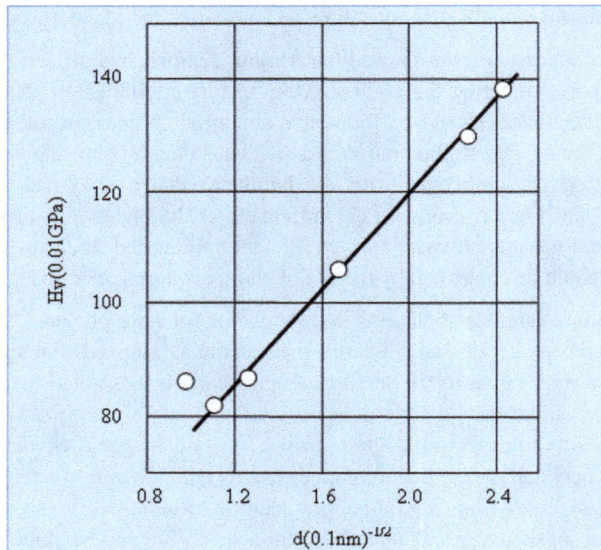

Figure 4. Variation with grain size of the microhardness of nanocrystalline silver

In other work[59], nanocrystalline samples were prepared via the *in situ* compaction of ultra-fine silver particles. The microstructure of the aggregates did not change when the compaction pressure was increased from 0.25 to 2.00GPa. The pressure affected the structure and density of the boundaries between the aggregates, such as the formation of approximately 1μm crack-type defects at the boundaries. The thermal stability was very low, with grain-coarsening beginning below 200C. On the other hand, a nanometre-sized layered structure formed in local regions during annealing and was stable at up to 800C. The Vickers microhardness of as-compacted specimens increased with increasing compacting pressure. This increase was attributed to a decrease in the number of crack-type defects. The microhardness began to decrease, due to grain-coarsening, upon annealing at about 200C. The number of crack-type defects in specimens compacted at 2.00GPa was lower than that in specimens compacted at 0.50GPa, but the standard deviation of data from a straight line for specimens compacted at 2.00GPa was greater than that for specimens compacted at 0.50GPa. The deviation was therefore not attributed to the effect of crack-type defects. It was concluded that the hardness of nanocrystalline silver inherently deviated from the Hall-Petch law.

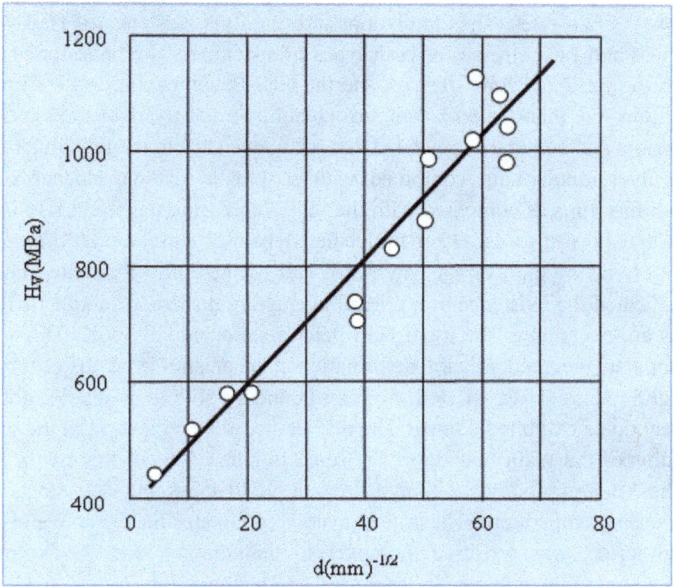

Figure 5. Hall-Petch relationship for nanocrystalline silver

Using the Tabor rule, the yield stress was given by

$$\sigma_y(MPa) = 123.71 + 3.29d(\mu m)^{-1/2}$$

When nanocrystalline material was prepared[60] by spark plasma sintering, using silver nanoparticles produced via the arc plasma method, the relative density and mean grain-size increased with increasing sintering temperature and time. The mechanical properties depended upon the relative density as well as upon the mean grain-size. The Vickers hardness of dense sintered samples obeyed the Hall-Petch law (figure 5). In tensile tests, sintered samples exhibited work-hardening and a uniform elongation, although the mean grain-size was less than 500nm. The yield strength of samples sintered at 600K for 60s was 5.2 times higher than that of coarse-grained samples. The relationship between yield strength and grain size also obeyed the Hall-Petch law.

Platinum

Large-scale molecular dynamics simulations have been used[61] to model the behaviour of nanocrystalline bulk and thin-film samples having average grain-sizes ranging from 5 to 40nm at two strain-rates. The simulations indicated the existence of Hall-Petch maxima for the yield and flow stresses of both types of specimen. The presence of free surfaces decreased the yield and flow stresses, and the Hall-Petch maximum for films occurred at a larger grain-size than for bulk samples. Quantitative analysis of the plastic slip within grain interiors and boundaries revealed that the above shift in the maximum resulted from a higher intergranular slip combined with a weaker size-dependence of dislocation activity within films as compared with the bulk. Increasing the strain-rate increased both the yield and flow stresses. Other molecular dynamics simulations[62] have been used to study the effects of the average grain size and temperature. Simulated uniaxial tensile results indicated the existence of a critical average grain size, of about 14.1nm, at which there was an inversion of the usual Hall-Petch relationship at 300K. This was attributed to a change in the predominant deformation mechanism: from dislocation motion, at average grain sizes above 14.1nm, to grain-boundary sliding at smaller grain-sizes. The Young's modulus exhibited a linear dependence upon the reciprocal of the grain size, and the modulus of the grain boundary was equal to some 42% of that of the grain core at 300K. The Young's modulus, ultimate strength, yield stress and flow stress all decreased with increasing temperature. The critical average grain size for the above inversion of the Hall-Petch effect was sensitive moreover to temperature, and the Young's modulus exhibited an approximately linear dependence on temperature.

Aluminium

In early work[63] on nanocrystalline aluminium, samples with an average diameter of 16.4nm were produced by radio-frequency magnetron sputtering. The grain size increased to 98.0nm during isothermal annealing at 573K. For grain-sizes ranging from 15 to 100nm, the hardness versus grain-size data were well represented by the Hall-Petch relationship. The coefficient of friction of aluminium against a stainless-steel pin varied with the sliding distance. In the early stages of sliding, the coefficient of friction rose to a peak value and then decreased to a steady-state value. The transition in the friction behaviour corresponded to a similar transition from severe to mild wear above a characteristic sliding distance on the cumulative volume-loss versus sliding-distance curve. The peak coefficient-of-friction decreased from 1.4 for material with a grain-size of 106nm to 0.6 for material with a grain-size of 16.4nm. The coefficient-of-friction of nanocrystalline material exhibited a 30% increase when tested in vacuum. In the nanocrystalline range, the wear rate was linearly dependent upon the square root of the grain-size, and this was described by using an empirical equation which was based upon Archard's Law.

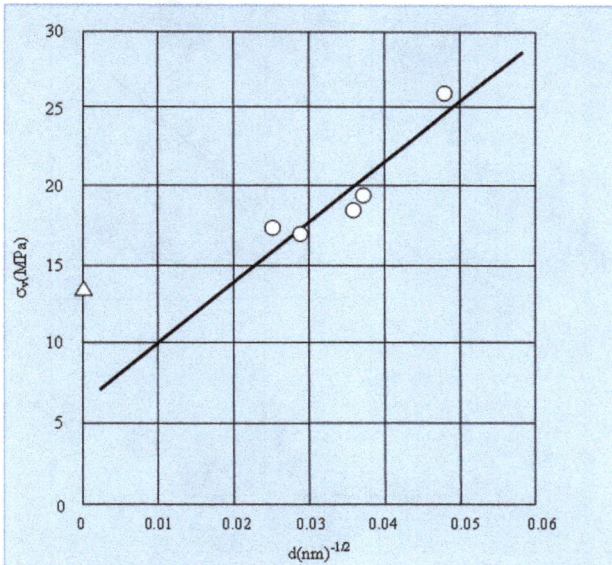

Figure 6. Hall-Petch plot of the yield stress of nanocrystalline aluminium. The triangle indicates the yield stress of coarse-grained material

Materials Research Forum LLC
https://doi.org/10.21741/9781644900352

Room-temperature uniaxial tensile testing[64] of nanocrystalline specimens, prepared by mechanical attrition and cold consolidation, with an average grain size of 20 to 40nm yielded stress-strain curves which revealed an increased tensile strength and a reduced ductility as compared with those of coarse-grained material. This strength-increase was much lower than that which was to be expected from extrapolation of the Hall-Petch law (figure 6) to nanometre-sized grains. The strengthening-rate was also strongly reduced in comparison with that of coarse-grained material.

Uniaxial tensile testing of 30 to 50nm-thick free-standing aluminium films showed[65] that the Young's modulus and ductility decreased monotonically with grain-size. Non-linear elasticity and slight irreversible deformation was observed in the 50nm-thick specimens. Inverse-Hall-Petch behavior was observed, with no appreciable room-temperature creep.

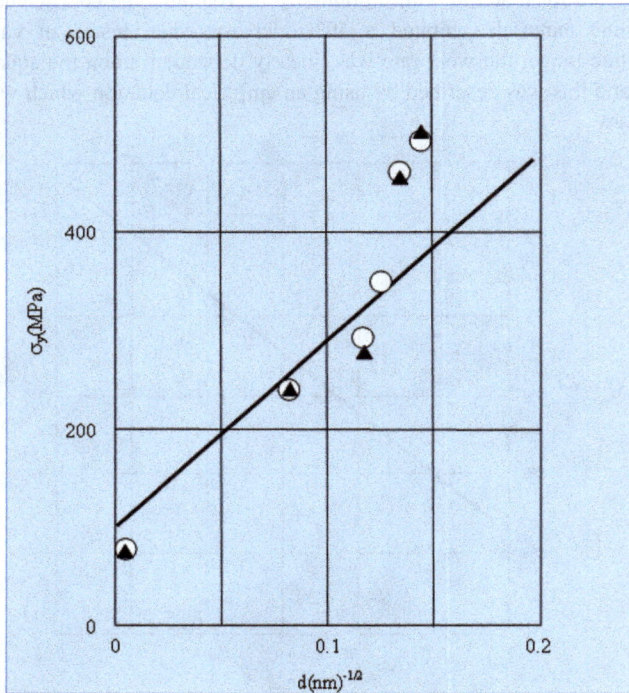

Figure 7. Grain-size dependence of the yield stress of aluminium in compression (circles) and tension (triangles)

Nanocrystalline particles having various sizes were prepared by ball-milling and consolidated into bulk specimens by hot pressing, before being subjected to quasi-static and dynamic compression[66]. As expected, a reduction in the grain-size of the strain-rate-dependent material led to a several-fold increase in the hardness and strength. On the basis of the results, the Khan-Huang-Liang viscoplastic model was modified by including a bi-linear Hall-Petch relationship. The modified equation then yielded results which were very close to experimental observations. The strain-rate-dependent responses for various grain-sizes were in good agreement with experimental results for grain-sizes ranging from the micrometric to the nanometric.

The deformation of bulk nanocrystalline material produced by the hot extrusion of ball-milled powder, and having grain-sizes as small as 48nm, was studied using compression and tension testing[67]. The isotropic yield stress increased, in line with the Hall-Petch relationship for grain-sizes down to 70nm. A positive deviation in the relationship occurred however at grain-sizes below 70nm (figure 7). Within the specific grain-size range of 60 to 48nm, structural analysis[68] indicated that perfect-dislocation emission governed the deformation behaviour because the inherent properties of aluminium of high stacking-fault energy and low twinability did not permit any other deformation mode. Samples having a grain-size of 48nm had a yield stress of 500MPa and exhibited ductile failure.

Severe plastic deformation was imposed[69] on pure aluminium specimens by means of high-pressure torsion. The samples consisted of A1050 (2N-purity), A1050 (2N5-purity), 4N-purity, 5N-purity and 6N-purity aluminium. The hardness of 6N-purity aluminium decreased with strain, and saturated at a level which was below that of annealed samples. This softening behavior was not observed in 5N-purity or lower-purity aluminium. The grain-size dependence of the hardness became less marked with increasing purity level, while high-pressure torsion-processed 6N-purity aluminium obeyed a reciprocal Hall-Petch law. In 6N-purity specimens having a large grain-size, dislocations accumulated in the grains in the form of dislocation cells and increased the hardness. When the grain size was small, the dislocations moved quickly and disappeared into high-angle grain boundaries.

When bulk nanocrystalline material with a thickness of 250µm was electrodeposited[70] from a dimethylsulfone bath it had a hardness of about 1.7GPa, and this was explained in terms of the Hall-Petch relationship and solid-solution strengthening mechanisms. That is, the grain size of some 40nm, and solid-solution strengthening by trace impurities, contributed to the above high hardness value.

Bulk material having a nano-scale grain-size was prepared[71] by high-energy ball-milling of pure aluminium powder. The average grain size ranged from 47 to 66nm after ball-milling, and the microhardness first increased and then decreased, before finally increasing again, with lengthening ball-milling time. The microhardness attained its maximum value when the ball-milling period was 5h. Meanwhile the yield strengths and grain-sizes obeyed the Hall-Petch law.

An analytical model was proposed[72], for predicting the tensile and compressive properties of bimodal nanocrystalline alloys, which relied upon simple material variables. It led to a Hall-Petch grain-size dependence, and included a new ductility parameter which could be used – together with the Hollomon equation - to predict the failure stress and strain. Application of the model to bimodal nanocrystalline aluminium alloys led to excellent agreement with experiment.

Having considered the pure metal at some length, how will its alloys behave. The structure of oxygen-doped aluminium films varied from columnar polycrystalline to nanocrystalline and to amorphous with increasing oxygen concentration, leading to markedly different mechanical properties[73]. The hardness increased rapidly, with increasing oxygen content, in agreement with the Hall-Petch relationship.

Nanocrystalline microstructures containing a high fraction of intermetallic phases were produced[74] in Al-Cu-Zr Al-Ni-Zr and Al-Si-Zr alloys by means of rapid solidification. The intermetallic phases and nano-scale grains led to hardness values of up to 5GPa in the case of melt-spun specimens. The presence of zirconium imparted a hardness which was at least twice that of the corresponding binary eutectic, and was some 5 times that of coarse-grained aluminium. An estimate of the various contributions to strength revealed that strengthening due to grain refinement was the predominant mechanism, and that the Hall-Petch effect predominated over Orowan strengthening by a factor of 4. This grain-size effect was greatly modified by the presence of grain-boundary precipitates.

Perhaps the newest class of metallic materials comprises the so-called high-entropy alloys which break every classical rule of successful alloy design. A single-phase face-centred cubic nanocrystalline solid-solution in equi-atomic AlCoCrCuFeNi was produced[75] by ball-milling. The milled powders had a plate-like morphology and a lattice parameter of 3.641Å. Compaction of the ball-milled powder into bulk specimens using spark plasma sintering at 1023K led to the precipitation of ordered body-centred cubic B2 phase. The sintered alloy microstructure exhibited a bimodal grain-size distribution, with average grain-sizes of 112 and 1550nm, plus face-centred cubic and B2 solid solutions, dislocations and twin boundaries. A hardness of 6.5GPa was found for samples sintered at 1023K for 0.25h. Consideration of the possible strengthening mechanisms suggested that frictional stress, Taylor hardening, Hall-Petch strengthening, solid-solution strengthening and twin-boundary strengthening were responsible. The Taylor hardening which arose from the intersection of dislocations and grain-boundaries, including Hall-Petch strengthening, accounted for 85% of the measured flow stress. The Tabor ratio (which generally states that the yield stress is equal to one third of the Vickers hardness) was up to 2.7; in close agreement with that for conventional polycrystalline materials. Nano-indentation using a force of 8000µN indicated a hardness of 8.13GPa and an elastic modulus of 172GPa. The strain-rate sensitivity was 0.0084 and the activation volume was $13b^3$. This suggested that grain boundaries, twin boundaries and interphase boundaries governed the deformation kinetics.

A study[76] of the solid-solution and grain-boundary hardening due to magnesium in Al-1.5Mn-0.5wt%Cu alloys containing 0.1 to 2.1wt%Mg, performed at 25 to 345C, revealed a transition at about 200C. Below this temperature, the frictional stress, the solid-solution hardening due to magnesium and the grain-boundary hardening were little affected by temperature. This was attributed to increased solute-pinning of dislocations by Cottrell clouds or other extended solute segregations. Above the transition temperature, solid-solution hardening was controlled by a temperature-dependent shear modulus, which decreased linearly with increasing temperature, up to about 300C. Both the frictional stress and the solution-hardening due to magnesium decreased linearly, with increasing temperature, above about 200C. The grain-boundary hardening obeyed the Hall-Petch law over the entire temperature range. Below the transition point, its slope was controlled mainly by magnesium solute. Above the transition, it depended upon both the magnesium concentration and the temperature. The Hall-Petch slope here increased linearly with magnesium concentration, and decreased in proportion to the reciprocal temperature.

As-cast ingots of manganese-modified cubic $L1_2$ titanium trialuminide, Al-25.6Ti-9.4at%Mn, were homogenized (1000C, 100h), crushed to a coarse powder and ball-milled for up to 386h in shearing mode[77]. The crystallite size remained essentially unchanged (20 to 30nm) during up to 100h of milling. It then suddenly decreased to a limiting size of about 3nm after 200h of milling. The lattice strains tended to be less than 1%. The ordered $L1_2$ structure began to disorder after about 40h of milling, and this ended after 160h of milling. With increasing milling time, the particles appeared to contain a work-hardened core, in which each particle was surrounded by a heavily deformed hard outer layer which contained nanometre-sized grains. After 386h of milling, all of the core and outer-layer particles had transformed into uniform core-less ones having a 3nm crystallite size. The microhardness data for the outer layer of particles with a core, and those for core-less particles, could be fitted by a Hall-Petch equation of the form,

$$H_V(kg/mm^2) = 431.7 + 387.5d(nm)^{-1/2}$$

Copper

In a classic early study[78] of the tensile behavior and microhardness of nanocrystalline copper as a function of grain size[79], the increase in strength with grain refinement continued down to the finest-grained samples (figure 8). A review of previous work[80] at that time suggested that the negative Hall-Petch slopes which were sometimes observed at ultra-fine grain-sizes were not associated with room-temperature creep but with the use of single samples which had been annealed repeatedly in order to change the grain size.

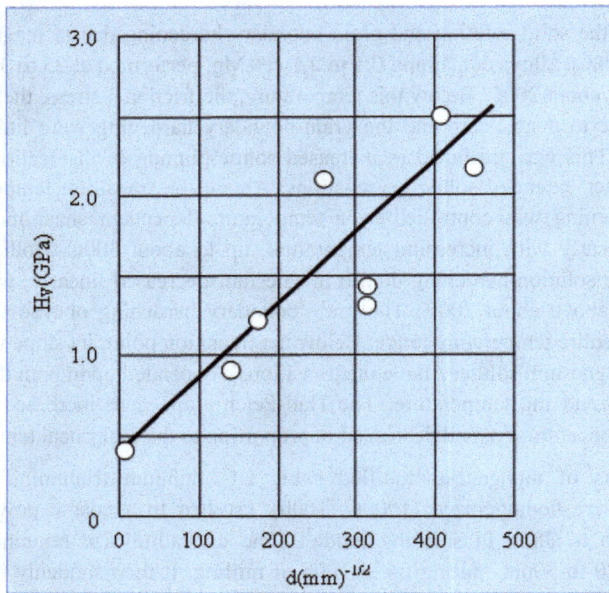

Figure 8. Microhardness of copper as a function of grain size

Nanocrystalline samples, consolidated from powder prepared by inert gas condensation, had grain-sizes ranging from 3 to 50nm, with lattice strains ranging from 0.02 to 3% and densities ranging 97 to 72% of that of coarse-grained material. The microhardness of the nanocrystalline samples exceeded that of annealed coarse-grained material by 2 to 5 times, in spite of the effect of porosity. The uniaxial tensile strength of nanocrystalline samples was also greater than that of coarse-grained material[81]. Restrictions on dislocation generation and mobility, which were imposed by the ultra-fine grain size were suggested to be the predominant factor in increasing the strength. Room-temperature diffusional creep, which was predicted to be considerable in nanocrystalline samples, was not observed although there were signs of a much smaller logarithmic creep.

In further work[82], samples having grain-sizes of 10 to 110nm and densities greater than 98% of theoretical were produced by inert-gas condensation and warm compaction. The yield strength of pure nanocrystalline material was 10 to 15 times higher than that of annealed coarse-grained specimens. Total elongations of 1 to 4% were observed in samples having grain-sizes less than 50nm, but a sample having a grain size of 110nm exhibited an elongation of more than 8%. Hardness measurements obeyed the Hall-Petch relationship down to about 5nm and then levelled off. The hardness values were 2 to 3 times greater than the tensile yield strengths. When compression tests were performed[83] on disks of nanocrystalline material, the yield strengths were of the order of 1GPa but were very dependent upon porosity; an appreciable fall in yield strength being associated with a 1 to 3% decrease in density. The compressive yield strength of 98%-dense copper having a grain-size of 20nm lay on a Hall-Petch curve which had been extrapolated from coarse-grain data. The nanocrystalline samples, although brittle in tension, underwent extensive compression before failure. An unusual mechanical behaviour of sub-micron grain copper was reported[84] in which there existed a combination of high strength, high ductility and lengthy stage of stable plastic flow at a very low strain-hardening rate. This was explained in terms of a non-equilibrium structure of the grain boundaries and triple junctions. Ultrafine-grained samples exhibited much higher strengths than did conventional polycrystalline copper and, as expected, the yield stress increased with decreasing grain size. It was noted here, perhaps for the first time, that nanocrystalline and sub-micron samples were weaker than was to be expected upon extrapolating the Hall-Petch relationship into the ultra fine-grained region. When the mechanical properties of three bulk ultrafine-grained materials, produced by equal channel angular pressing, were studied[85] by using strain-rates ranging from 0.001 to about 4500/s, the temperature and strain-rate sensitivity of all three was appreciably higher than that which was typical of annealed face-centred cubic polycrystalline metals. The yield strengths here tended to lie above those extrapolated from other Hall-Petch data, suggesting that sub-structures, in addition to the average grain-size, might be important in influencing the yield strength.

When the plastic deformation of nanocrystalline copper, with and without 0.2wt% of boron, was studied[86] using fully-dense defect-free compacts, the Hall-Petch relationship was obeyed down to the smallest grain-size investigated (25nm) and the Hall-Petch slopes were lower for nanocrystalline than for conventional copper.

Thermomechanical treatment of copper resulted[87] in a bimodal grain-size distribution, with micron-sized grains embedded within a matrix of nanocrystalline and ultra-fine (<300nm) grains. The matrix grains imparted the high strength which was to be expected from extrapolating the Hall-Petch relationship, while an inhomogeneous microstructure induced a strain-hardening mechanisms which stabilized tensile deformation; leading to high tensile ductility, a 65% elongation-to-failure and a 30% uniform elongation.

A copper matrix which contained niobium filaments was compared[88] with a copper matrix which contained niobium tubes filled with copper. When the characteristic dimension of the structure was greater than 10μm, there was no size-effect of the composite hardness. Within the 1 to 10μm range, a marked increase in hardness reflected a change in the plasticity mechanism, and was attributed to the classical Hall-Petch effect. In the nanometric range, the hardness of the nanocomposite regions exceeded that of nanocrystalline copper or niobium, and could attain 5.8GPa. The observed size-effect on the plasticity of Cu/Nb nanostructures, together with the dislocation-barrier role played by Cu/Nb interfaces, confirmed models which were based upon the occurrence of a single dislocation regime at nanometric scales.

Compaction of micron-sized copper powder produced[89] nanocrystalline specimens having a density of 99.4% and a mean grain-size of 34 to 43nm, with a micro-strain of 0.16 to 0.19%. The grain-size of nanocrystalline samples could be correlated with the compacting pressure, and the microhardness was 1.14 to 1.27GPa. This was about two times higher than that of coarse-grained polycrystalline copper. The relationship between the microhardness and the grain size did not strictly obey the Hall-Petch law.

The hardness of nanocrystalline samples having grain-sizes as small as 10nm continued to obey the Hall-Petch relationship[90]. A rate-sensitivity of 0.06 and a flow-stress activation volume of $8b^3$ were deduced for a grain size of 10nm. This suggested that grain-boundary activity was increased, but did not yet predominate during plastic deformation.

Twinning becomes a major deformation mechanism in nanocrystalline copper during high-pressure torsion using very low strain-rates at room temperature. It has been suggested[91] that many twins and stacking faults form in nanocrystalline copper via partial dislocation emission from grain boundaries. Results showed that the Hall-Petch relationship could break down in nanocrystalline copper because of a change in deformation mechanism.

A study[92] was made of a composite which consisted of alternating closely-spaced nanocrystalline nickel and copper films, electrodeposited onto cold-rolled polycrystalline copper substrates. Vickers microhardness testing, using loads ranging from 1.96 to

0.049N, had to take account of the fact that, above a certain critical penetration, the hardness was not that of the electrodeposited film but was a so-called composite hardness. The microhardness increased, with decreasing layer-thickness down to 30nm, in accord with the Hall-Petch relationship.

The effect of coherent twin-boundary and stacking-fault spacings upon the deformation of nanowires was investigated[93] by using molecular dynamics simulations, revealing that there was a marked change in the mechanical behavior of nano-twinned wires when the twin-boundary spacing was less than 10.23nm. There was thus an optimum twin-boundary spacing, and the peak stress decreased with decreasing coherent twin-boundary spacing. For specimens having various stacking-fault spacings, the peak stress exhibited a similar trend to that in the case of nano-twinned specimens. When the stacking-fault spacing was sufficiently fine, the stacking-fault acted as a strong barrier. The deformation behavior of material containing alternating parallel twin-boundaries and stacking-faults was also investigated, and the results suggested that that would lead to a higher peak stress and potentially high peak strain.

The deformation behavior of compressed electrodeposited nanocrystalline copper pillars, having grain-sizes of 360, 100 or 34nm, was investigated[94] as a function of the specimen size. In recent years, two relevant length-scales have emerged in the quest for ultra high-strength polycrystalline metals. The microstructural length-scale, of grain-size or twin-size, has traditionally been linked to the familiar Hall-Petch relationship, but the sample length-scale, the nanopillar size, has now proved to be as relevant, particularly in micro-scale structures. The yield stress of pillars having grain-sizes of 360 or 100nm did not depend upon the specimen size. The material instead exhibited roughly the bulk yield stress until the specimen size was reduced to a critical specimen-size/grain-size ratio of 35 or 85. The yield stress then decreased with decreasing specimen size. The yield stress for pillars with a size of 34nm hardly depended upon the specimen size, and exhibited essentially the bulk yield stress for all of the specimen sizes investigated. The predominant deformation mechanism changed from dislocation glide, for pillars with a size of 360 and 100nm, to grain-boundary diffusional creep for pillars with a size of 34nm. A size-induced softening occurred for pillars with a size of 34nm, in accord with the change in deformation mechanism. The bulk yield stress for pillars with a size of 360 or 100nm increased in accord with the Hall-Petch relationship. The critical specimen-size/grain-size ratios, for pillars having grain-sizes of 360 or 100nm, increased with decreasing grain size, in accord with the power-law scaling found for coarse-grained copper polycrystals. This suggested that specimen-size induced softening could extend from the micron to the nanometre range if the predominant deformation mechanism was dislocation glide. The large critical ratios found for pillars with grain-sizes of 360 or

100nm were explained in terms of strain-continuity among neighboring grains, and the generation of geometrically necessary dislocations; required in order to maintain strain-continuity at the grain-boundaries.

The behaviour of monocrystalline and polycrystalline copper under spherical nano-indentation was investigated[95] by means of molecular dynamics simulation. In order to study the grain-boundary network effect, the grain-boundary interface was sub-divided into three microstructural components having differing dimensions. The indentation force for monocrystalline material was larger than that for polycrystalline material, and the latter force continuously decreased, with decreasing grain size, due to a softening phenomenon. Defect nucleation and propagation occurred below the tool tip, in both monocrystalline and polycrystalline copper, due to the high internal stresses and atomic potential energy which were introduced by nano-indentation. The horizontal propagation of defects was faster and more extensive than their vertical propagation, and the defects were restricted to the grains around the tool tip due to the grain-boundary network. A clear stress and potential-energy gradient existed beneath the tool tip in monocrystalline copper, while the gradients could be distributed among multiple directions in polycrystalline samples.

Powders of Cu-Ta alloy were synthesized[96] using high-energy cryogenic mechanical alloying and consolidated into bulk nanostructured specimens by means of equal channel angular extrusion at high temperatures. The microstructures consisted of an equiaxed copper matrix containing fine tantalum precipitates whose size distribution varied with composition and processing temperature. In some cases, there was an almost three-fold increase in mechanical properties over those predicted by Hall-Petch estimates for pure nanocrystalline copper. An attempt[97] was made to obtain nanocrystalline bulk Cu-40at%Fe alloy by mechanical alloying, followed by consolidation and annealing. The consolidation permitted supersaturated solid solutions to precipitate Cu_4Fe and $CuFe_4$. Limited grain growth occurred following annealing. Hardening increased during up to 40h of milling, and the variation in microhardness as a function of grain-size revealed a Hall-Petch behavior. Copper alloys having niobium concentrations ranging from 1 to 10at% were studied[98] following annealing at 400 or 800C. Grain-growth inhibition at both temperatures increased with increasing niobium content. Following annealing at 800C, Cu-1Nb, Cu-5Nb and Cu-10Nb alloys had grain-sizes that were about 8, 14 and 14 times smaller, respectively, than that of unalloyed copper. This inhibition was attributed to the formation of niobium oxide based clusters, niobium segregation zones and large niobium-based precipitates. Following annealing at 400 or 800C, Cu-5Nb possessed a hardness which was some 2.5 times or 3 times, respectively, that of the hardness of unalloyed copper following the same annealing treatment. The increase was attributed to Hall-Petch and precipitation strengthening.

With decreasing grain-size a possible change in deformation behaviour, from that of nanocrystals to that of metallic glasses, can be postulated. This possibility was modelled[99] using molecular dynamics simulations of bulk glasses, of glass–crystal nanocomposites and of $Cu_{64}Zr_{36}$ nanocrystals containing various crystalline volume fractions. The grain-boundary phase was found to behave like a metallic glass, constrained by the adjacent crystallites. The change from glass-like to grain-boundary controlled plasticity was divided into three stages. At low crystalline volume fractions, the system was essentially a glass–crystal composite and plastic flow was localised in the amorphous phase. With increasing crystalline volume fraction, clusters of crystallites became jammed together and the mechanical response depended greatly upon relaxation of the glassy grain boundaries. At grain-sizes above 10nm the system became completely jammed, thus preventing pure grain-boundary plasticity and leading instead to co-deformation. An inverse-Hall–Petch effect occurred only during the second stage, when the grain boundaries were not deeply relaxed.

An immiscible nanostructured Cu-Al-Nb alloy was prepared[100] by friction stir processing of mechanically compacted pellets and aged at 563K for various times. The material exhibited a hardness of about 4.3GPa in the peak-aged (6h) condition, and this excellent strength was attributed to Hall-Petch strengthening arising from the extremely refined nanocrystalline structure, together with precipitation-strengthening by a uniform distribution of nano-scale aluminium- and niobium-rich precipitates or clusters. Nanocrystalline $Cu_{86}Al_{12}B_2$ alloy having an as-milled average grain-size of about 11nm was synthesized[101] by high-energy ball-milling at cryogenic temperatures before annealing at up to 900C; preserving its nanoscale grain size. First-principles calculations revealed a preferred segregation of boron to the grain boundaries of Cu-Al alloy. A high thermal stability was attributed mainly to grain-boundary pinning of $CuAl_2$ and AlB_{12}, and the alloy maintained a high hardness. Grain-size dependent Hall–Petch strengthening was the predominant mechanism, together with a large degree of solid-solution and Orowan strengthening in the annealed state.

The effect of the grain-size upon the flow-stress of face-centred cubic polycrystals was analyzed[102] by using a multi-scale approach which was based upon a mathematical homogenization of the grain aggregate. The behavior of each crystal was assumed to be governed by a dislocation-based plasticity model with a Taylor-model critical resolved shear stress. The generation and annihilation of dislocations in each slip system was assumed to follow a Kocks-Mecking model, as modified so as to account for dislocation-storage in the grain boundaries. Polycrystalline copper was chosen as a test-bed for the simulation, and the model parameters were derived from dislocation-dynamics simulations or experimental data for smaller length-scales. It was found that the initial dislocation density played the dominant role in deciding the magnitude of the grain-size effect, and that the 'reciprocal power of grain-size' dependence of the flow-stress broke down in the case of initial dislocation densities which were greater than $10^{14}/m^2$ and for grain-sizes which were greater than 40µm. The grain-size contribution to the strength obeyed a power-law function of the initial dislocation-density when the applied strain was less than 2%.

Nickel

A secant-viscosity composite model was developed[103] in order to explain the strain-rate sensitivity of nanocrystalline solids. The composite was assumed to consist of a grain-interior phase and a grain-boundary affected zone, with the properties being described using a viscoplastic constitutive law. The drag stress of the grain interior was proposed to obey the Hall-Petch law, while that of the grain-boundary affected zone was deemed to be independent of grain size. In terms of the secant viscosity of the above phases, the strain-rate sensitivity was then determined using a linearly viscous comparison composite and a field-fluctuation approach. The results indicated that the tensile strength of nanocrystalline nickel with a grain-size of 40nm should be some 5 times that of microcrystalline material having a grain size of 10µm, at the same strain-rate (3×10^{-4}/s). The nanocrystalline material was also expected to exhibit a much higher strain-rate effect. Possible grain-size softening with further grain-size reduction was also predicted.

The Hall-Petch relationship was investigated[104] in a nano-twinned alloy with no dislocation pile-ups. This showed that, when the twin spacing was greater than 150nm, the hardness exhibited a reciprocal square-root dependence on the spacing. When the twin spacing was less than 100nm, the dependence was of simple reciprocal form. These results were explained in terms of dislocation-controlled mechanisms. The effect of grain orientation upon the mechanical behaviour of electrodeposited nanostructured nickel was such that the influence of grain-size upon flow stress was accurately described by the Hall-Petch law for a wide range of grain-sizes greater than 20nm, but the Hall-Petch slope depended upon the texture orientation with respect to the distribution of grain boundaries[105]. These observations were explained in terms a competition between grain-boundary shearing and dislocation emission at grain boundaries.

When nanocrystalline thin films were produced[106] by electrodeposition using sulfamate solution, 3 types of film could be distinguished: constant-current, pulsed-current and constant-current with grain-refinement additive. The grain-size became smaller in that order. The tensile fracture and yield strengths obeyed the Hall-Petch law, and the elongation was greatest for constant-current with grain-refinement additive films. The fatigue strength increased with decreasing grain size, again obeying the Hall-Petch law down to about 10nm. The resistance to fatigue-crack propagation decreased for nanograin films. The threshold stress intensity factor was smallest for pulsed-current and constant-current with grain-refinement additive films. In the intermediate-rate propagation range, the rate increased with decreasing grain size for a given stress intensity factor. The fatigue-fracture surface near to the threshold consisted of granular features, the size of which decreased with decreasing grain size. At high stress-intensity factors, striations were found on the fracture surface of constant-current films, while only fine granular features were observed for constant-current with grain-refinement additive films. On specimen surfaces near to the fatigue-fracture surface of pulsed-current and constant-current with grain-refinement additive films, small grain-boundary fracture-facets were observed at high stress-intensity factors, but were not observed near to the threshold. In constant-current films, slip bands were seen together with grain-boundary fracture-facets. In related work, nanocrystalline films were prepared[107] by pulse electrodeposition onto steel substrates, from an additive-free sulfamate-type bath. Increasing the peak current density was again more effective in improving the mechanical response than was varying the pulse-on time. All of the deposits had a preferential orientation and the effect of grain-size confirmed the Hall-Petch law. Thin films having various grain-sizes were also produced[108] by electrodeposition, at two bath temperatures, from sulfamate solutions which contained various brightener contents. The grain-sizes which were deduced by Fourier analysis of 111-222 diffraction pairs were smaller than those for 200-400 diffraction pairs; for films with a 200 texture. The opposite was found for films possessing a random orientation. The distribution of grain-sizes followed a log-normal distribution. The mean grain size, as determined by X-ray diffraction of 111-222 pairs, agreed with that determined by transmission electron microscopy, for grain-sizes below 30nm, but was smaller for larger sizes. The size which was deduced using Scherrer's method and 111 diffraction was close to that found via Fourier analysis of 111-222 pairs, and agreed with transmission electron microscopic results for sizes that were lower than 30nm. The grain-size decreased with increasing brightener content at the lower bath temperature. The yield stress increased with decreasing grain-size, and almost obeyed the Hall-Petch law down to a grain-size of 20nm. Pulsed electrodeposition onto AA6061 substrates, from a modified Watts bath with saccharin grain-refining additive, produced[109] nanocrystalline samples having average grain-sizes ranging from 115 to 17nm. The hardness increased as the grain-size decreased, reaching 7.2GPa for a coating with a grain-size of 17nm. The elastic modulus remained almost constant, at between 150 and 160GPa, regardless of the average grain-size. The coefficient-of-friction was 0.25 for a grain-size of 17nm. No inverse-Hall-Petch relationship was observed however within this

Materials Research Forum LLC
https://doi.org/10.21741/9781644900352

range of grain-sizes. A nanocrystalline pure nickel coating was electroplated onto pure copper by using a modified Watts bath and various stirring-rates and current densities[110]. The crystallite size exhibited a reciprocal dependence upon the stirring-rate and current density. The best corrosion resistance was found for a $5A/dm^2$ current density. The evolution, at high stirring-rates, of surface bubbles affected the surface quality and properties. The Vickers microhardness increased with increasing current density, but exhibited a reciprocal dependence upon the degree of turbulence in the solution. Optimum agitation of the bath could improve the hardness distribution of coated specimens, but those specimens did not obey the Hall-Petch relationship. The electrodeposition of nanocrystalline coatings from baths containing deep eutectic solvents was studied[111]. This effective addition of extra water led to a more uniform and finer-grained deposit. The nickel coatings had an average crystallite size of about 5 to 7nm. The increase in the water content of the plating bath resulted in an increase in the microhardness of the deposits, and an inverse-Hall-Petch effect was observed.

An atomistic study was made[112] of the plastic deformation of columnar nanocrystalline nickel structures, with regard to dislocation emission from grain boundaries. The samples comprised grain boundaries having random tilt misorientations arranged around a common <110>-type axis, and all of them consisted of a 36-grain polycrystalline microstructure with grain-sizes ranging from 4 to 20nm. Tensile deformations of up to 8% were simulated, and the stress-strain curves exhibited grain-size effects in both the elastic and plastic ranges. An inverse-Hall-Petch effect occurred at a nominal stress, for a fixed strain, but disappeared when a grain-size dependent elastic modulus was used to construct an 0.5%-offset yield stress. Dislocation-emission from the grain boundaries, and grain-boundary accommodation of plasticity, both took place. The dislocation-emission occurred mainly from pre-existing dislocation-like structures in grain boundaries and increased rapidly at grain-sizes greater than 4nm. The number of dislocations per unit length of grain boundary stayed at a constant value for large grain sizes, indicating that there was a fixed density of pre-existing sources in the grain boundaries. Preventing dislocation-emission within a short distance from the triple-junctions reproduced the overall density versus grain-size relationship. Grain-boundary sliding occurred in the same regions of the microstructure for all grain-sizes, and to about the same extent. Such a simple model for dislocation and sliding phenomena was consistent with observed trends in plastic strain as a function of grain-size.

During a comparison[113] of the tensile behavior of hexagonal close-packed nanocrystalline cobalt and face-centered cubic nanocrystalline nickel, having grain-sizes of about 20 and 28nm, respectively, it was noted that their tensile ductility, elongation-to-failure and fracture behaviour were very different. More pertinently, it was noted that there was considerable scatter in the data for nickel, thus making it difficult to judge which Hall-Petch law should be adopted. Earlier work on the yield strength of nanocrystalline nickel had usually been deduced from hardness measurements of small samples which had exhibited almost brittle behaviour and a low tensile strength. Other data resulted from tensile tests. Unfortunately, two very different Hall-Petch fits to the data were possible (figure 9):

$$\sigma_y(GPa) = 0.23GPa + 13.33(GPa/nm^{-1/2})\, d(nm)^{-1/2}$$

$$\sigma_y(GPa) = 0.078GPa + 6.49(GPa/nm^{-1/2})\, d(nm)^{-1/2}$$

It was concluded that, because nickel is a face-centred cubic metal with an anomalously small friction stress (table 1), the second equation was the more reasonable one.

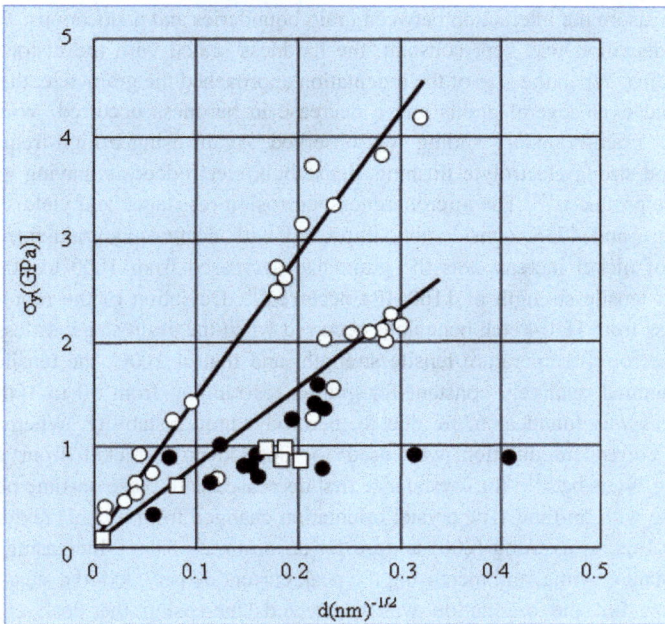

Figure 9. Possible Hall-Petch laws for nickel
Open circles: hardness tests, filled circles: tensile tests, squares: present work

Table 1. Hall-Petch constants for various metals

Metal	Friction Stress GPa)	Slope (GPa/nm$^{-1/2}$)
Nickel	0.078	6.49
Copper	0.0255	3.478
Cobalt	0.432	1.90
Titanium	0.0785	12.65

Nanocrystalline material was produced[114] by pulsed electrolysis and was heat-treated so as to give grain-sizes which ranged over the nanoscale to 30nm. Nano-indentation was used to measure the interaction between grain boundaries and dislocations. When the size of the indentation was kept constant, the hardness scaled with the dislocation density within grains. When the size of the indentation approached the grain-size, the plastic zone was spread over several grains and a decrease in hardness occurred. With decreasing grain-size, grain-boundary sliding was observed. Again using a high-frequency pulsed current and strong electrolyte flushing, dense bulk electrodeposits having various grain-sizes were produced[115]. The microhardness, corrosion resistance and yield stress at room temperature and 473K considerably improved with decreasing grain-size. The tensile strength of nickel increased as the grain-size decreased from 1000 to 70nm, where a maximum tensile strength of 1160MPa occurred[116]. Deviation of the room-temperature yield stress from Hall-Petch behaviour occurred when the grain size was less than 70nm. Unlike the room-temperature tensile strength, and that at 200C, the tensile strength at 400C remained relatively constant for grain-sizes ranging from 50 to 1000nm. A low yield stress was found at 673K due to thermodynamic instability. When square-wave cathodic current modulation was used to electrodeposit nickel from a saccharin-containing Watts bath[117], the crystal size first decreased with pulse on-time before starting to increase with on-time. The crystal orientation changed from a (111) texture for an on-time of 0.1ms, to a strong (200) texture for an on-time of 8ms. Increasing the off-time with constant on-time, and increasing the peak current density, led to a steady increase in crystal size but the orientation was unaffected. Increasing the peak current-density markedly refined the crystal size. The orientation changed from almost random at the lowest peak current-density (0.2A/m^2) to a strong (200) texture at a peak current density of 2.0A/m^2. Nanocrystalline nickel having a grain size of the order of 30nm could be produced when using the above bath chemistry. The microhardness of the deposit was such that, when the grain size was large, it obeyed the Hall-Petch law but, when the grains were ultra-fine, there was a deviation from that law. When nanocrystalline nickel coatings were prepared using direct-current electrodeposition, the $C_7H_4NO_3SNa\bullet2H_2O$ (saccharin sodium) content, temperature and current density had a marked effect upon the electrodeposition rate and microhardness. The microhardness was higher for higher current densities, lower temperatures or higher $C_7H_4NO_3SNa\bullet2H_2O$ contents, and remained stable for current densities ranging from 700 to $1000A/m^2$. The relationship between mean grain size and microhardness roughly fitted the Hall-Petch law. In other studies it was found[118] that, with increasing $C_7H_4NO_3SNa\bullet2H_2O$ concentration, the grain size first decreased and then increased, with the smallest grain-size corresponding to a concentration of 10g/l. The microhardness was also a maximum for that concentration, as was the corrosion resistance. The microhardness again roughly obeyed the Hall-Petch law.

Sub-micron columnar nickel structures having various cross-sectional geometries and grain-sizes were prepared[119] using electron-beam lithography and electroplating. Uniaxial micro-compression showed that the flow stress increased at smaller grain sizes as usual. A size-dependent softening effect was observed only in nickel pillars having grain-sizes ranging from 9.4 to 13.2nm. On the other hand, pillars having grain-sizes close to 22nm did not exhibit any flow-stress dependence upon size; the pillar strength remaining essentially constant. There was no dependence of the flow stress upon the cross-sectional geometry for pillars having a given microstructure.

It was shown[120] that the grain-size of the (111) crystal plane decreased as a function of the texture coefficient, regardless of the processing conditions, while that of the (220) plane behaved differently. The dependence of the hardness upon the grain-size was closely described by the Hall-Petch relationship when changing the temperature or pH was involved, albeit with various slopes. It deviated from the relationship upon changing the current density, suggesting that other underlying mechanisms, related to texture, were occurring. There was a notable degree of influence of the texture upon the hardness and the elastic modulus, with the hardness tending to increase with texture. When the texture coefficient of (111) was greater than 40%, the elastic modulus increased with texture, implying that there was a fundamental relationship between the modulus and the texture, with the latter outweighing any other factors.

When nanocrystalline NiAl was instead synthesized[121] by electron-beam gas-condensation and compacted *in situ* at various temperatures, the as-compacted specimen microstructures were stable at up to about 1000C and had grain-sizes ranging from 2 to 4nm. The microhardness increased with an increasing density which was due mainly to a reduced porosity. The microhardness also increased with increasing grain-size following annealing, contrary to be behaviour of coarse-grained NiAl. The material was thus markedly stronger than the coarse-grained form but was not as strong as an extrapolated

Hall-Petch law predicted. The nanocrystalline form also exhibited room-temperature ductility, unlike the coarse-grained material. It was proposed that diffusion, rather than dislocation, mechanisms controlled the strength and ductility. On the other hand, when nanophase powders were prepared via the mechanical alloying of elemental nickel and aluminium powders under argon for up to 48h and consolidated by shock-compaction at 4 to 6GPa, the average crystallite size ranged from 48 to 9nm but microhardness measurements revealed only normal Hall-Petch behaviour[122].

A study[123], at between ambient and 300C, of nanocrystalline nickel which contained carbon concentrations of 500 to 1000ppm revealed increasing grain growth with increasing temperature; and particularly rapid growth at 300C. Pure nanocrystalline nickel exhibited abnormal grain growth at 500C, and its tensile properties were markedly impaired, so the addition of carbon was useful in improving the stability of the microstructure at intermediate temperatures. On the other hand, carbon-doped nickel had inferior tensile properties, but they obeyed the normal Hall-Petch law. Bulk nanocrystalline nickel having a carbon content of 30 to 1600ppm and a sulfur content of 140 to 1200ppm was prepared[124] by using a sulfamate electrodeposition bath containing various agents. The hardness data were in accord with the Hall-Petch relationship as the grain size decreased to about 12nm but were scattered with increasing carbon and sulfur contents; the scatter being attributed to the effect of impurities. In order to separate out the various contributing factors to the hardness, the solution enthalpies and misfit strains were calculated using first-principles methods. The results indicated that carbon existed as an interstitial solute atom in the matrix, producing large misfit strains, while sulfur existed as a substitutional solute atom and had essentially no effect. The solution strengthening was attributed to interstitial solute atoms interacting with mobile dislocations.

Nanocrystalline Ni-Co deposits were prepared[125] by pulse-plating, using constant electrodeposition conditions but varying concentrations of saccharin and cobalt sulfate. Suitable amounts of the additions produced a finer structure and a greater hardness, although further additions could result in a decreased hardness; an effect which was attributed to the inverse-Hall-Petch phenomenon. The maximum hardness of the Ni-Co alloy deposits was no greater than that of pure nickel control samples, suggesting that the refinement-hardening (Hall-Petch) effect predominated in nanocrystalline Ni-Co alloy deposits. When alloy coatings which contained 0 to 45wt%Co were electrodeposited[126] from a modified Watts bath, the morphology and grain-size of the coatings were greatly affected by the cobalt content: with increasing cobalt content, the surface morphology changed from pyramidal to spherical. The microhardness increased with increasing cobalt content; due mainly to a decreasing grain-size, in accord with the Hall-Petch effect.

Bulk nanocrystalline γ-phase alloys containing 19 to 21wt%Fe were prepared[127] using mechanochemical and hot-isostatic pressing methods, giving yield strengths – for a grain-size of 33nm – which were some 13 times greater than that of the coarse-grained alloy. The change in yield strength with grain-size was roughly in accord with the Hall-Petch law, for grain-sizes ranging from 33 to 100nm. Nickel-iron nanocrystalline samples having various compositions and grain-sizes were prepared[128] by electrodeposition, with the iron content being chosen by varying the Ni/Fe ion-ratio of the electrolyte. For a constant grain-size of about 11nm, the hardness strongly depended upon the iron content, with a maximum at around 20%Fe. The grain-size of specimens containing 4 to 6%Fe was very sensitive to the deposition conditions, while those containing 35%Fe or more cracked upon deposition and the tendency to crack was related to the level of internal stress in the deposit. The hardness of these alloys obeyed the normal Hall-Petch relationship. The thermal stability of electrodeposited nickel-21%Fe having an average grain-size of some 14nm was investigated[129] by annealing (373 to 773K, 1.5h). Grain growth began after annealing at 400K or above, and two ranges of grain growth were noted. In the range below 575K, growth was accommodated largely by grain-boundary diffusion. At higher temperatures, it was assisted by lattice diffusion. The microhardness results reflected the existence of the two ranges. The positions of the (111) X-ray diffraction peaks indicated a sudden decrease in the lattice parameter upon annealing at 773K, and this was attributed to a loss of connectivity of the nanosized grains. The hardness again obeyed the Hall-Petch law. Bulk nanocrystalline samples of nickel and Ni-15wt%Fe were obtained[130] via electrodeposition, and the grain-size of nickel samples was varied from 15 to 200nm by changing the deposition parameters. The grain-size was further reduced to 9nm by alloying with iron. The strength and strain-hardening rate increased with decreasing grain size, and the fracture behavior depended upon the grain-size, the presence of defects and the stress state. The tensile elongation and reduction-in-area varied greatly from sample to sample and were not related to the fracture behavior. Various types of behavior were noted: in some cases, the samples exhibited entirely ductile fracture but very little elongation. In other cases the samples exhibited relatively brittle behavior but a marked tensile elongation. In a third case, the samples exhibited ductile behavior and a reasonable tensile elongation.

When C276 (Ni-Mo-Cr) layers were deposited[131] onto 4140 steel substrates by plasma-enhanced magnetron sputtering, the as-deposited 12μm-thick coating had a face-centered cubic nanocrystalline structure, with a grain size of 37 to 46nm, and its hardness of 6.2GPa was 4 times higher than the values of 1.5 to 1.9GPa which were reported for the bulk material. This difference was attributed to the small grain-size and hence to the Hall-Petch effect. Annealing (1h, 600 to 800C, air) had little effect upon the grain-size, but further increased the hardness to between 7.3 and 8.5GPa; an increase which was attributed to segregation and to the precipitation of chromium-rich and molybdenum-rich material.

The effect of solutionising treatment upon the microstructural evolution and mechanical behavior of Ni-Mo-Cr alloy, N10276, was such that the average grain-size remained almost unaffected by holding at 1050C. This was attributed to pinning of the grain boundaries by secondary phases. A sharp increase in grain-size and grain growth-rate occurred upon increasing the solutionizing temperature to 1150C; regardless of the holding time. This was attributed mainly to the complete dissolution of precipitated particles and to the increased diffusivity of solutes. The rate of grain growth decreased with extended solutionizing time and decreasing temperature. The dissolution of second-phase particles, together with grain growth, increased the work-hardening exponent and markedly improved the tensile ductility but generally reduced the tensile strength. Applying the Hall–Petch function to the experimental data on samples solutionized above 1150C, yielded a Hall–Petch coefficient of 18.35MPa√mm for the yield strength[132].

Inverse Hall-Petch Effect

The hopes that further and further reduction in the grain-size of nanocrystalline material would lead to unprecedented levels of material strength were unfortunately dashed when it was found that the extrapolation was unfounded. That is the gist of the problem: why does it break down, and how can it be prevented?

The normal Hall-Petch equation, linking strength to the reciprocal of the square root of the grain-size, annoyingly breaks down at grain sizes which are of the order of some tenths of a nanometre. One immediate conclusion drawn was that this was due to a gradual switch from intragrain-controlled deformation to boundary-controlled deformation. Such a transition has been shown to be feasible by means of finite-element simulations, although this is not conclusive because only a few grains, rather than the millions which are implicated in reality, tend to be considered. Although it is well known that the grain size has a statistical distribution in polycrystals and nanocrystals, this also tends not to be considered in the analysis of deformation. This too might exert an effect on the transition from grain-boundary strengthening to grain-boundary weakening in nanocrystalline materials. That transition is expected to broaden with an increase in the standard deviation of the grain-size distribution.

Nanocrystalline materials are defined by their characteristic microstructural length of up to approximately 100nm. Those having grain sizes of some 0.1 to 0.3µm are termed sub-micron materials. When the grain size is less than 10 to 20nm, it becomes a geometrical inevitability that more than 50vol% of the atoms will be associated with the various boundaries. In this situation, it is argued that dislocation pile-ups cannot form and the Hall-Petch law, which was specifically predicated on the existence of such pile-ups is consequently then invalid. But given that the grain boundaries then play the major role in deformation, the latter process becomes subject to the effects of phenomena such as creep, superplasticity and diffusion-mediated grain-boundary sliding and rotation.

Theoretical Models

Various methods have been used to modify the original theory of the Hall and Petch era so as to deal with nanocrystalline material.

A new model[133] for the Hall-Petch relationship, in nanocrystalline material, was proposed in which the normal relationship was well-described by a grain-boundary dislocation-source model. A transition from normal to abnormal Hall-Petch behaviour was predicted to occur at a point where the grain size was smaller than the average spacing between the grain-boundary sources. The abnormal Hall-Petch behaviour could then be described by,

$$H_v\text{-}H_m = \beta\sqrt{(1\text{-}3\delta_{eff}/d)}$$

where H_v and H_m were the hardnesses of the nanocrystalline and amorphous forms of the material, respectively, δ_{eff} was the effective grain boundary thickness, d was the grain size of the nanocrystalline material and β was a constant. As mentioned above, the distribution of the grain sizes played an important role in demarcating the transition region. Also reconsidered[134] was the role played by dislocation pile-ups in determining the degree of Hall-Petch strengthening of a material. At the nanoscale, certain features of the pile-ups such as a paucity of dislocations and elastic anisotropy could have a marked effect. Use[135] of the bow-out dislocation line model led to a modified Hall-Petch relationship for nanocrystalline materials in which the predicted,

$$\tau/G \sim d^{-1/2}$$

curve was in good agreement with the trend in experimental,

$$H_v/G\alpha \sim d^{-1/2}$$

relationships for a number of nanocrystalline materials, where α was a constant converting shear-stresses to hardness values. At large grain sizes, the present relationships again became the usual Hall-Petch relationship.

Taking into account the above anisotropy of crystallographic symmetry and various possible choices of critical shear strength led[136] to a reasonable limit on the grain sizes for which dislocation pile-up theory could be applied to nanocrystalline materials. Deviation from the Hall-Petch relationship was explained in terms of a dislocation pile-up mechanism involving small numbers, and a composite model was proposed to explain the strength of nanocrystalline materials. This model could rationalize the various cases which were observed in Hall-Petch studies. An expression for incorporating the creep-rate of nanocrystalline material, as governed by a diffusion mechanism, was developed which also included the effect of triple-line diffusion. Triple junctions, the line of intersection of three or more grain boundaries, are inherent microstructural elements of polycrystalline materials and possess unique characteristics which depend upon the crystallographic orientation of the adjoining crystal lattices, as well as exhibiting distinct energies, corrosion susceptibilities, ductilities and diffusivities. It will be found that the creep-rate due to triple line diffusion exhibits a greater grain-size dependence than does that due to grain-boundary diffusion. In view of this the standard mechanisms of creep, via Nabarro-Herring diffusion through grains or Coble creep involving grain boundaries, were modified[137] so as to allow for transport along triple-lines. Transitions between the various diffusion-driven creep processes were determined, and it was demonstrated that triple-line diffusion creep was likely to be significant in nanocrystalline materials. Unlike normal materials, nanocrystalline ones having a grain size that was below a critical value were not expected to exhibit a transition from diffusion creep, at low stresses, to conventional intragranular power-law dislocation creep at high stresses. This was attributed to the difficulty of creating dislocation pile-ups in such fine-grained materials. The increased degree of mass transport which was associated with triple-line diffusion was predicted to contribute markedly to an increased superplastic deformation.

One model[138] for the plastic deformation of nanocrystalline materials was couched in terms of the evolution of a disclination-grid, located at the triple junctions of grains. Plastic deformation was assumed to take place as the result of plastic rotation of the grains; the mismatch of their rotations then provoking the nucleation of partial disclinations at the junctions of intergrain boundaries[139]. A process which delayed an increase in the capacity of Frank vectors at the plastic-flow stage was an ongoing mechanism which included the emission, absorption and re-emission of dislocations by disclinations.

It has been shown[140] that quasi-periodic tilt boundaries in nanostructured polycrystals are capable of contributing to deviations from the usual Hall-Petch dependence of yield-stress upon grain-size. The behavior of a collection of interacting grain-boundary defects in polycrystals was analyzed[141] by using a statistical approach showing that, as the grain-size decreased, the defect system underwent a topological transition from soliton-like shear-instability waves to spatially periodic defect structures. This led to marked changes in the spatial and temporal evolution of the defect collection. The appearance of the nanocrystalline state was directly related to the development of periodic defect structures, and equations were deduced which explained deviations from the Hall-Petch relationship and a transition from Herring-Nabarro to Coble diffusion.

Poor sample-quality can easily lead to large deviations from the mechanical behavior expected of flaw-free material. The compression of high-density nanocrystalline metals led[142] to high hardness and yield-strength values which were consistent with the extrapolation of coarse-grained Hall-Petch data into the nanocrystalline regime.

Given that mechanisms such as the glide of lattice dislocations and grain-boundary phenomena may be active during the plastic deformation of polycrystals having a small grain-size, a continuum model was developed[143] which encompassed such features. A strain-gradient effect due to dislocation pile-ups, dislocation emission or absorption at the surface and grain-boundary were accounted for in terms of energy storage. The model was applied to a bicrystal under plane-constrained shear, and the dependence of the yield strength upon the bicrystal thickness was deduced. It was predicted that the yield strength of the bicrystal would first increase and then decrease with decreasing thickness; thus recalling the inverse-Hall-Petch effect, but with different scaling. The behaviour was attributed to changes in the predominant deformation mechanism which occurred at various specimen thicknesses.

The assumption of differing behaviours of the bulk and grain-boundary regions is routinely invoked in order to explain the special features of nanocrystalline materials, such as the arrangement of geometrically necessary dislocations and disclinations. The usual Hall-Petch law is assumed to describe the hardness-dependence of each of the two regions and a corrective term is introduced for the Hall-Petch slope of the boundary region as compared to that for the grain-interior slope. An overall formula for the total hardness was obtained[144] here which involved the grain size, the grain-boundary width, the hardness of coarser-grained materials plus a parameter which described the arrangement of the triple junctions.

A constitutive equation which predicted the effect of grain-size upon the yield stress of metals, and which was based upon the model of Meyers and Ashworth, was extrapolated[145] into the nanocrystalline regime. At large grain sizes, it had a Hall-Petch form while, in the nanocrystalline domain, the slope gradually decreased and asymptotically approached the flow stress of the grain boundaries. It was surmised[146] that 4 main factors contributed to grain-boundary strengthening[147]. These were the fact that grain boundaries acted as barriers to plastic flow, that they acted as dislocation sources, that elastic anisotropy generated additional stresses in the vicinity of grain boundaries and that multi-slip is initiated in grain boundaries whereas the grain interiors are initially ruled by single-slip when properly oriented. Consequently the regions which abut grain boundaries harden at a rate which is much higher than that of the grain interior.

Computational predictions were made of plastic flow as a function of grain size incorporating differences of dislocation accumulation rate in grain-boundary regions and grain interiors. The material is modelled as a monocrystalline core surrounded by a mantle (grain-boundary region) with a high work hardening rate response. This is the first computational plasticity calculation that accounts for grain size effects in a physically-based manner.

In another model[148], the dislocation glide contribution to the plastic strain rate was proposed to vanish below a critical grain size. The grain-boundary material, treated as a separate phase, was assumed to deform via a diffusional mechanism, leading to viscous Newtonian behaviour. The strain in both phases was assumed to be the same and to be equal to the imposed macroscopic strain. The stress was then calculated by using a simple rule-of-mixtures argument. The resultant grain-size dependence of the stress-strain curves, and the predicted strain-rate effects, were in reasonable agreement with experiment. Observed deviations from Hall-Petch behaviour were correctly described by the model.

The uniaxial tensile deformation of computer-generated nanocrystalline samples having average grain-sizes ranging from 5.38 to 1.79nm was simulated[149] by means of molecular dynamics, assuming a Finnis-Sinclair potential. The results revealed a reverse Hall-Petch effect. Grain-boundary sliding and motion, plus grain-rotation were deemed to be largely responsible for the plastic deformation. At low temperatures, partial dislocation activity played a minor role during deformation. This began at a strain of 5%, and increased sharply with increasing average grain size. At high temperatures, no dislocation activity was detected, and diffusion of grain boundaries was suggested to occur.

The question naturally arises, when modelling grain-size related deformation processes, as to how many grains should be included. The inverse-Hall-Petch has been widely attributed to a gradual switch from intragrain-controlled deformation to grain-boundary controlled deformation; a transition which has been observed in many finite-element simulations in spite of the practical restrictions which limit the choice of the assembly to just a few grains, rather than millions. This limitation is generally overlooked or ignored. The approximation was quantified[150] by considering a finite number of grains and applying a simple analytical model. It transpires that the finite-element approximation is relatively good, even when the sample is only 3 grains in width. The results can moreover be adjusted so as to correct the stress-strain curves of small representative volumes. A question also arises as to exactly what shape the grain has. The mean grain-size and

Materials Research Forum LLC
https://doi.org/10.21741/9781644900352

dispersion are generally used as parameters for designing nanomaterials having the required mechanical properties, but grain-size here implies a putative equivalent radius which assumes that the grains are spherical. Two grains having the same size may nevertheless have different shapes, and may consequently exert differing mechanical effects upon neighbouring grains. An attempt was made[151] to predict nanomaterial properties on the basis of observed features of the grains. It was tested by using data on nanocrystalline aluminium. The relationships between the statistics of microscopically observed cumulative number of grain faces, the grain density and mechanical properties were analogous to the Hall-Petch to inverse-Hall-Petch relationship change.

When the stress distribution near to grain boundaries was considered in order to improve the dislocation pile-up model for the Hall-Petch relationship, the continuous distribution of dislocations in the pile-up could be effectively handled[152] by using Tschebychev polynomials for the Hilbert transformation, and an analytical formula could be deduced for the stress intensity factor of the dislocation pile-up. The inverse-Hall-Petch relation could also be explained by the modified dislocation pile-up-model.

In a somewhat odd mathematical analysis[153], a recent attempt has been made to explain the inverse-Hall-Petch effect by using fractal theory. Assuming little more than packings of various numbers of small circles within the same larger circle, it was concluded that the fractal dimension is the fundamental criterion. That is, when the fractal dimension is greater than unity, the normal Hall-Petch effect occurs. When the fractal dimension is smaller than unity, the inverse-Hall-Petch effect occurs.

With regard to the atomistic simulation of the strength of solids, various theoretical definitions of strength have been essayed. When defined by modes of structural instability they are, in the long-wavelength limit, specified by criteria which involve elastic stiffness coefficients and the applied stress[154]. In more general terms, strength can be characterized by the onset of soft vibrational modes in the deformed lattice. On the other hand, the molecular dynamics simulation of stress-strain responses yields a direct measure of the effects of small-scale microstructure upon strength. A Hall-Petch type of scaling can be introduced in order to estimate the strength of specimens which contain microstructural flaws of a certain critical size.

In a further refinement[155], a self-consistent scheme was used to describe the behavior of nanocrystalline face-centred cubic materials in which the latter was approximated by a two-phase composite where the inclusions were the grain cores while the matrix represented both the grain boundaries and the triple junctions. Dislocation glide was incorporated into the constitutive law of the inclusions, while a thermally activated

Materials Research Forum LLC
https://doi.org/10.21741/9781644900352

mechanism which accounted for the penetration of dislocations into the grain boundaries was incorporated into the constitutive law of the matrix.

The strength of polycrystalline materials was summarised[156] by mapping (figure 10) the deformation mechanisms of nanocrystalline materials using molecular dynamics simulations. The deformation map demarcated three regions of deformation behavior, depending upon the grain-size and stress-level; both being normalized with respect to the stacking-fault energy. Dislocation emission could be blocked within the confined geometry of nanocrystalline grains when the stacking-fault energy was sufficiently low.

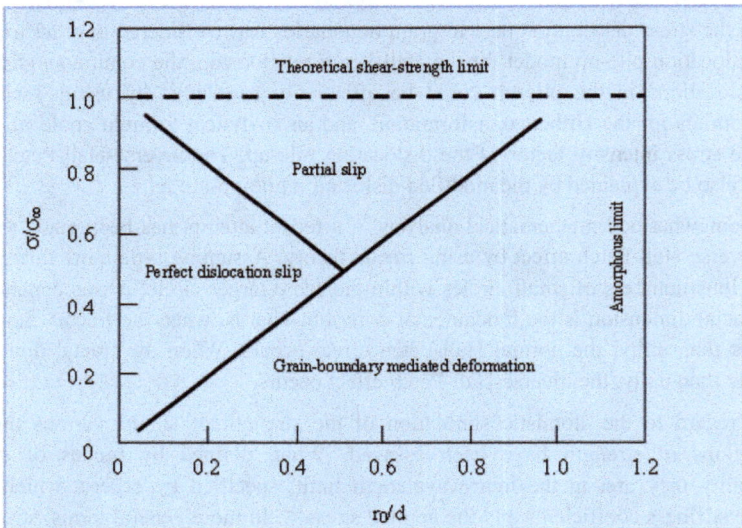

Figure 10. Mechanism-map (reduced-stress versus reciprocal grain-size) for the effect of stacking-fault energy upon deformation behaviour: distinct regions in which either complete extended dislocations, partial dislocations or no dislocations exist during the low-temperature deformation of nanocrystalline face-centered cubic metals.

It was noted[157] that nanocrystalline metals are sub-divided by grain boundaries while heavily-deformed metals are sub-divided by a mixture of dislocation boundaries and high-angle boundaries, and their characteristic dimensions range over 5 length-scales; from the nanometric to the hundreds of micrometers. A multi-scale analysis was therefore applied to the relationship between strength and boundary parameters. Such an analysis

Materials Research Forum LLC
https://doi.org/10.21741/9781644900352

suggested that the yield-stress versus grain-size relationship obeyed the Hall-Petch law down to a grain-size of some 15 to 20nm.

It was pointed out[158] that the strain-rate sensitivity of face-centred cubic metals was increased, in the nanocrystalline range, due to the decrease in activation volume. This was not observed in body-centred cubic metals because the activation volume was already low in the conventional range. In fatigue tests, the stress-cycle curves seemed to show some improvement due to the increase in strength, but the rate of crack advance increased; due possibly to the fact that a smoother fracture required less energy to propagate. Creep results were ambivalent, with some results indicating a decreased creep resistance that was consistent with a small grain-size while others showed that the creep resistance was not negatively affected. An increased tendency to twinning resulted from the increased separation between partial dislocations. The fracture surfaces of nanocrystalline metals consisted of a mixture of ductile dimples and sheared regions. The dimple-size was much smaller than that of conventional polycrystalline metals, but was several times larger than the grain size. The sheared regions were a direct result of the increased tendency of nanocrystalline metals to exhibit shear-localization.

Mechanical data on nanocrystalline materials indicate the existence of a critical grain size such that, at larger sizes, the strength increases with decreasing grain size while, for smaller sizes, the strength decreases with decreasing grain-size. By combining the conventional Hall-Petch law for larger grains with a new deformation process which was based upon dislocation-accommodated boundary-sliding for smaller grains, it was shown[159] that nanoscale softening could be predicted. The critical grain sizes for nanocrystalline copper and nickel were found to be 25 and 13nm, respectively. Such findings will be considered in greater detail later.

Given that, in nanocrystalline solids, significant numbers of atoms are located within the grain boundary and the adjoining outer grain, the combined region – the grain-boundary affected zone - is softer than the grain interior. The combined behaviours of the grain interior and of the grain-boundary affected zone then govern the overall response. A two-phase composite model has been used[160], to model the high-temperature creep resistance of nanocrystalline material, in which the rate-equation of each phase was of power-law type and the grain interior was taken to obey the Hall-Petch law while the grain-boundary affected zone response was grain-size independent. The overall constitutive equation was also consistent with the concept of secant viscosity, in that a transition from linear visco-elasticity to non-linear viscoplasticity was involved, with the Maxwell viscosity being replaced by the secant viscosity. A field-fluctuation method was then used to determine the effective stress on each phase. Application of the theory to the creep resistance following grain-size decreases in the nanometre range showed that the creep resistance in a Hall-Petch like plot underwent a transition from a positive slope to a horizontal one, and then to a negative slope. The horizontal value indicated the maximum creep resistance that a material could attain, and this occurred at a critical grain size that was in the nanometre range. Preparation methods for nanocrystalline solids, such as ball milling, may lead to the presence of voids in the final product which can impair its yield strength. An analytical composite model[161] was able to explain the viscoplastic response of porous nanocrystalline material, and conceived the latter as comprising three phases: a plastically harder grain interior, a plastically weaker grain-boundary affected zone plus pores. Use of the same homogenization and fluctuation methods as those used in the previous two-phase model then linked the local strain-rates to the total applied strain-rate. That link again provided the secant viscosity of the constituent phases during each deformation stage. It was found that this model was able to replicate the main features of experimental data, given the existence of various grain sizes and porosities. The calculations again predicted a change in the yield strength of the Hall-Petch plot: from an initial increase, to a plateau, and then to a decline, for various porosities under constant strain-rates[162]. There was again a critical grain-size, in the nanometric range, at which the material exhibited a maximum yield strength. This critical grain size tended to shift to the left, in the Hall-Petch plot, as the grain-boundary affected zone became softer.

Some results have suggested[163] that, when the grains of polycrystalline material are refined into the nanocrystalline range, there exist two critical sizes at which there occur changes in the hardening mechanism. A generalized equation was deduced for describing the dependence of yield strength upon grain size for a wide range of parameters with regard to the changes in the nature of the dependence as the critical grain sizes were attained. It has also been noted[164] that there are scaling constraints associated with the plastic deformation of nanograined polycrystals, with regard to the analysis of data on their mechanical properties. In particular, it has been suggested that the systematic transfer to nanograined metals, of concepts successfully applied to coarse-grained equivalents, has no firm basis.

Another preparation method, nanocrystalline electrodeposition, is applicable to a wide range of pure metals, metallic alloys and metal-matrix composites. Here again, the strengthening effect resulting from grain-size reduction is governed by the Hall Petch effect[165].

A dislocation-density grain-boundary interaction scheme, a grain-boundary misorientation dependent dislocation-density relationship, and a grain-boundary sliding mechanism were combined[166] in order to model the behavior of nanocrystalline material having a grain size ranging from 25 to 200nm. The above features were coupled to a dislocation-density multiple-slip crystalline plasticity formulation and a finite-element algorithm so as to predict the response of a nanocrystalline aggregate. Account was taken of slip-system compatibility, local resolved shear stresses and changes in immobile and mobile dislocation densities. A dislocation-density conservation law was used to balance dislocation-density absorption, transmission and grain-boundary emission. The predicted relationship between yield stress and grain size was consistent with the Hall-Petch law. It was also deduced that grain-boundary sliding and grain-size effects affected crack behaviour via local dislocation-density and slip evolution at critical grain boundaries. Most importantly in the present context, it was predicted that grain-boundary sliding increased with decreasing grain size and led to lower normal stresses existing in critical locations. Grain-boundary sliding could therefore counter the strength increase which was expected to be associated with a decrease in grain size.

The fatigue behaviour of nanocrystalline metals offers some clues as to various dislocation mechanisms which might be relevant to their Hall-Petch strengthening. Early research on the fatigue behavior of nanocrystalline metals revealed evidence for an improved fatigue resistance as compared to that of normal microcrystalline metals. Experimental and modelling observations revealed the details of cyclic plasticity, microstructural stability, crack-initiation and crack-propagation processes. In cyclic plasticity studies, nanocrystalline metals exhibit a strongly rate-dependent cyclic hardening, thus reflecting the importance of diffusive deformation mechanisms such as grain-boundary sliding. Cyclic deformation also causes substantial mechanically-induced grain-coarsening, recalling that observed during the high-strain monotonic deformation of nanocrystalline metals. Crack-initiation in nanocrystalline metals is associated with sub-surface internal defects and with surface extrusions, but it remains unclear how extrusions form when the grain size is below that required for persistent slip-band formation. Nanocrystalline metals offer very little resistance to crack propagation, due to a limited plasticity and to a lack of crack-path tortuosity. Like bulk metallic glasses, nanocrystalline metals nevertheless exhibit ductile fatigue striations and a metal-like Paris-law behavior.

It has been noted that a polycrystal will transform into a glass if the grain-size is sufficiently reduced, but it is not clear where the demarcation occurs and whether it is sharp or diffuse. Experimentation in this domain being difficult, simulations were used[167] to compress binary monocrystals, comprising hard and soft disks, first into polycrystals and then into glasses. This made it possible to increase continuously the fraction of mismatched particles, and produce minute grains. This wide range of grain-sizes then permitted study of the polycrystal-glass transition, and of the structural, mechanical and thermodynamic features at the sharp polycrystal/glass boundary. This simulation also naturally revealed the occurrence of Hall-Petch and inverse-Hall-Petch behaviors in the two-dimensional polycrystals. Three glass regimes were also identified. Compressing binary crystals could produce stable ultra-fine grains only when compressible portions were common and two suitable length-scales co-existed.

Partial-dislocation models for the deformation of nanocrystalline materials were extended[168] so as to include the effect of non-uniform dislocation extension. Such a model was more consistent with experimental data than was the original partial-dislocation approach.

Modelling of the relationship between yield stress and grain size at the nanoscale suggested[169] that the deformation behaviour could be demarcated into two regions: grain-rotation at the boundary and deformation in the inner ordered region. It was predicted that the yield stress would attain its maximum when the grain size was 3 times the grain-boundary thickness; corresponding to a 70% volume ratio of grain boundary. When the grain size was more than 3 times the grain-boundary thickness, the interface area increased with grain-size reduction; thus leading to a yield stress increase. When the grain size was less than 3 times the grain-boundary thickness, the interface area decreased with grain-size reduction, leading to a decrease in yield stress. The hardness exhibited similar trends.

Another dislocation model[170] was based upon the idea that dislocations which were emitted from grain-boundaries and which bowed-out into the grain interiors were responsible for plastic deformation and thermally-activated de-pinning at grain boundaries; thus making them rate-controlling.

A classical discrete dislocation dynamics simulation approach was extended[171] so as to account for grain-boundary sliding and for the absorption, emission and transmission of lattice dislocations at grain boundaries. The grain-boundary dislocations were assumed to nucleate, and migrate along the grain boundary, in a manner which was an extension of the original discrete dislocation dynamics formalism. The incorporation of a dislocation model for grain-boundary dynamics permitted all of the mechanisms to be able to relax localized stress-fields such as those resulting from dislocation pile-ups and to modify the mechanical response of polycrystals in a way which went well beyond the concept of grain boundaries simply blocking dislocation-slip.

Slip transfer at grain boundaries, and the interaction between propagating cracks and grain boundaries was analysed[172], with the emphasis being placed upon twin boundaries. It was shown that, in some situations, this transfer leads to normal motion of the twin boundary due to the displacement of partial dislocations along that boundary. Such motion could generate a de-twinning effect. The cleavage crack-path behaviour in body-centred cubic metals depended largely upon the twist component of the grain boundary.

The mechanism of propagation of such twisted cracks involved segmentation of the crack-front and the existence of intergranular parts. A similar segmentation appeared to occur for short fatigue cracks.

Other clues to dislocation and grain-boundary behaviour which might be relevant to modelling the Hall-Petch law at nanograin scales can be gleaned from the effect of irradiation. At the grain level, a tensorial crystal model[173] which incorporated irradiation and grain-size effects could be applied to the grain interior, while grain-boundary sliding with irradiation effects and grain-boundary diffusion were invoked when modelling the phenomena occurring at grain boundaries. An elastic-viscoplastic self-consistent method which took account of the grain-size distribution was used to link the microscopic behaviour of individual grains to the macroscopic properties of nanocrystals. Numerical results showed that irradiation-induced defects could lead to irradiation-hardening of the grain interiors, but the degree of hardening decreased with grain-size due to the increasing absorption of defects by grain boundaries. Absorbed defects could make the grain boundaries softer than in the non-irradiated case. There existed a critical grain-size, for irradiated nanocrystalline metals, which separated the grain size into an irradiation-hardening dominated region, at above the critical size, and an irradiation-softening dominated region when below the critical size. The distribution of grain sizes had an appreciable effect upon the mechanical behaviours of irradiated and non-irradiated nanocrystals.

A statistical grain boundary dislocation source model was proposed[174] for discrete dislocation-slip events. A grain-size limitation on dislocation source sizes gave rise to grain-size effects upon the statistical distribution for the critical resolved shear stress for discrete slip events. This grain-boundary source model was incorporated into a three-dimensional crystal-plasticity finite-element model. A Hall-Petch scaling of the yield strength arose from the calculations, and the predictions were in quantitative agreement with experimental data for a wide range of average nanoscale grain-sizes. A statistical dispersion of the critical resolved shear stress for discrete slip events led to strain-hardening of the macroscopic flow-stress versus strain response. The fraction of grains which accommodated the applied strain, and the fraction of active grains which underwent multi-slip, depended strongly upon the grain size. It also led to an unusual textural effect upon slip activity which was significant for finer nanograins but was weak for grain sizes greater than 100nm. There was also an increase in the heterogeneity of strain concentrations with decreasing grain-size. This suggested that plastic instabilities were more likely to occur as the nanoscale grain-size decreased.

It has been shown[175] that nano-indentation, of microstructures ranging from monocrystalline to nanocrystalline, can be used to deduce Hall-Petch parameters and flow curves by using just four pyramidal tips having various apex-angles.

The grain-size effect upon the elastoplastic behavior of ultra fine-grained and nanocrystalline materials has recently been modelled[176] by using a self-consistent approach and two different length-scales: the microscopic (crystallographic slip) and mesoscopic (grain/matrix interaction). The model could describe not only ultra fine-grained materials, with grain-sizes ranging from 100 to 1000nm, but also nanocrystalline materials having a grain-size of 100nm. This was characterised by a lower slope of the linear Hall-Petch relationship, as compared to that for the ultra fine-grained regime. Nanocrystalline metals having a grain-size (less than 25nm) which caused them to exhibit a plateau or a decrease in yield strength could not be described by the model.

The dependence of the yield strength upon the grain-size and upon the latter's dispersion may be different for quasi-static and dynamic loading conditions. The size-distribution of the grains affects the slope of the Hall-Petch curve differently in the two cases. Thus the existence of a wide size-distribution shifts the maximum yield strength to smaller grain sizes and may obscure any observation of an inverse-Hall-Petch effect. It is suggested[177] that the role played by the grain-boundary structure is significant at low strain-rates in coarse-grained material and at high strain-rates in fine-grained material.

Early geometrical structural models[178] of triple junctions showed that these defects could be the predominant structural elements of nanostructured materials. Experimental studies showed that mechanical behaviour, solute segregation and hydrogen transport could be accounted for by triple lines; especially the inverse Hall-Petch effect and enhanced diffusivity. It was postulated that triple-line defects represented a distinct structural link between polycrystalline, nanostructured and non-crystalline solids.

One model[179] for the deformation of nanocrystalline materials was based upon grain-boundary sliding and the formation of mesoscopic glide-planes. An experimentally observed decrease in hardness with decreasing grain-size, the inverse Hall-Petch effect, occurred in metals and intermetallics having nanometre grain sizes and could be described by the present model. In metallic systems a transition, from dislocation-controlled plastic deformation to grain-boundary sliding, occurred at a critical grain-size when the stress which was required for dislocation motion or formation became greater than that required for grain-boundary sliding.

Another model[180] specifically described the mechanical behaviour of nanocrystalline materials which were obtained via the crystallization of an amorphous precursor. The structure was treated as being a composite which consisted of an amorphous matrix and absolutely rigid inclusions which represented a crystalline phase. It was found that the yield-stress dependence exhibited a point of inflection at a critical grain size which was of the order of 20 to 25nm and an inverse Hall-Petch relationship was observed at grain sizes which were smaller than the critical value. This model further predicted the formation of a superlattice from disclinations which were located in the triple junctions of grains during the plastic flow stage of nanocrystalline material.

A polycrystalline model[181] for the abnormal grain-size dependence treated the nanocrystalline material as though it were a coherent-precipitate strengthened two-phase alloy within which all of the grain boundaries merged so as create an entire continuous matrix; each of the grains being coherently embedded in the matrix. According to this model, the transition from the normal to the inverse Hall-Petch relationship corresponded to an exchange of the roles played by the grain bulk and the grain boundary during deformation.

A theoretical investigation[182], based upon using the isothermal equation of state to explain the inverse Hall-Petch relationship and the expected limit of mechanical stability of grain boundaries in nanocrystalline metals, suggested that the excess free volume of grain-boundary atoms could induce softening and thus explain the phenomena which appeared below a critical grain size.

Nanocrystalline thin films of copper and zinc, having grain-sizes of 5 to 100nm, were found to exhibit hardness values which initially increased with decreasing grain-size according to the normal Hall-Petch relationship[183]. There was a critical grain-size below which the hardness decreased with decreasing grain-size. This seems to have been the first experimental evidence offered for the softening of nanocrystalline materials at very small grain-sizes; now known as the inverse-Hall-Petch effect. Most of the plastic deformation was envisaged here as being due to a large number of small sliding events which were associated with grain-boundary shear or grain-boundary sliding.

Assuming that, at grain sizes smaller than 20 to 30nm, dislocations play no significant role in the deformation of nanocrystalline materials, molecular dynamics simulations demonstrated[184] that - in the absence of grain growth and dislocations - nanocrystalline face-centred cubic metals could deform via a mechanism which involved a complex interplay between grain-boundary sliding and grain-boundary diffusion. By reproducing the Coble-creep formula for coarse-grained materials, it was shown that the inverse Hall-Petch effect arose from sliding-accommodated grain-boundary diffusion creep; thus unifying previously contradictory ideas based upon those mechanisms separately. A transition, with decreasing grain size, from a dislocation-based to a grain-boundary-based deformation mechanism, was further shown[185] to be associated with a marked change in the mechanical behaviour. That is, at the grain size at which the transition occurred, the strain-rate under tensile elongation went through a minimum. Such a simultaneous change in the deformation mechanism and in the corresponding mechanical behaviour could explain the change from Hall-Petch to inverse Hall-Petch behaviour.

It was noted[186] that the grain-size distribution could have an appreciable effect upon mechanical properties; especially in the case of grains of nanometre-size where the normal Hall-Petch law is replaced by the inverse-Hall-Petch law. A statistical model was proposed, for the relationship between flow stress and grain-size distribution, which considered various deformation mechanisms and was applied to the mechanical properties of aluminium and copper. This showed that the grain-size distribution played an important role in determining mechanical properties. The dependence of mechanical properties upon the grain-size dispersion also exhibited Hall-Petch and inverse-Hall-Petch behaviours. It was predicted that copper would be more sensitive to changes in grain-size distribution than would aluminium.

It was further revealed[187] by molecular dynamics simulation that, in the absence of grain growth, nanocrystalline molybdenum and UO_2 also deform via diffusion creep. In the case of molybdenum however both grain-boundary and lattice diffusion contribute to the creep rate. That is, the deformation mechanism involves a combination of Coble and Nabarro-Herring creep.

Inspired by such molecular dynamic simulations, a generalized self-consistent model was proposed[188] for studying changes in the yield-strength of polycrystalline metals as the grain-size decreased from the conventional coarse range and into the nanometre range. The simulations revealed that an appreciable fraction of the atoms was present in the grain boundaries, and that plastic flow of the grain-boundary region governed the mechanical characteristics. The model imagined that each oriented grain, plus its immediate grain boundary, constituted a pair which was in turn embedded in an infinite effective medium possessing properties which represented the orientational average of all of the pairs. The non-linear behavior of the nanocrystalline polycrystal was predicted by using the concept of secant moduli and the eigenstrain field of Luo and Weng. The plastic flow of each grain was deduced from its crystallographic slip, while the plastic behavior of the grain-boundary phase was modelled as though it were amorphous. The yield strength reached a maximum at a critical grain-size and the Hall-Petch slope then became negative in the nano-range. When Hall-Petch behaviour occurred, the plastic behavior was governed by crystallographic slips in the grains. When the slope was negative, it was governed by the grain boundaries. Within the transition range, both the grains and the grain boundaries contributed.

Materials Research Forum LLC
https://doi.org/10.21741/9781644900352

This approach was later reformulated[189] using an incremental small-strain scheme. The nanocrystalline material was again modelled as grain-plus-boundary pairs, embedded in an infinite effective medium. Plastic deformation of the inclusion-phase took account of the dislocation-glide mechanism, while the boundary phase was taken to be amorphous. The aggregate comprised spherical randomly-distributed grains having a grain-size distribution of log-normal type.

The consequences of the change in deformation mode were investigated[190] for systems that were subjected to large strains during cyclic deformation. In most coarse-grained metals, severe plastic deformation led to grain refinement, but simulations indicated that this process was suppressed in sufficiently small grains; instead, sliding in the grain boundaries markedly increased diffusive processes and led to grain-coarsening during deformation.

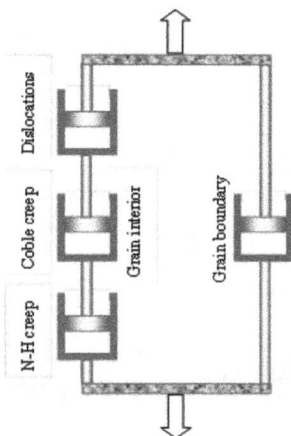

Figure 11. Voigt-type model for nanocrystalline deformation

A phase-mixture model[191] was used to simulate the deformation of metallic materials having grain sizes ranging from the micrometer to the nanometre scale. The polycrystalline material was treated as being a mixture of two phases. One was a grain-interior material, the plastic deformation of which was governed by dislocation and diffusion mechanisms. The other was a so-called grain-boundary phase, the plastic flow of which was controlled by boundary diffusion (figure 11). Very strain-rate sensitive behavior was observed for a strain-rate jump from 10^{-5} to 10^{-4}/s for a grain-size of 10 or

40nm, and was attributed to a change in the deformation mechanism from diffusion-control at low strain-rates to dislocation-glide control at high strain-rates. The strain-rate sensitivity was not so high for strain-rate jumps from higher strain-rate levels: being only 0.189 and 0.256 for grain-sizes of 10 and 40nm, respectively. At strain-rates above 0.1/s, the strain-rate sensitivity was always very low because the main deformation mechanism was dislocation glide. The strain-rate sensitivity increased with grain refinement at strain-rates of less than 10^{-3}/s and reached almost unity for grain-sizes of less than 20nm, due to predominantly diffusion-controlled deformation. The usual Hall-Petch behavior of coarse-grained material at high strain-rates, which was governed by dislocation glide, was replaced by inverse Hall-Petch behavior in ultra fine-grained materials at low strain-rates; provided that both phases deformed under mainly diffusion-control. A deformation-mechanism map (figure 12) captured the domains of predominance of dislocation-glide and diffusion-controlled plastic flow.

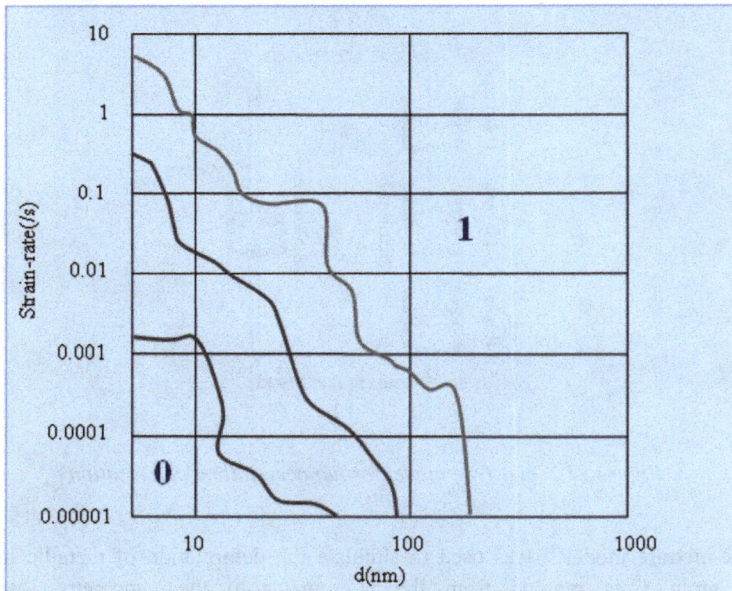

Figure 12. Deformation-mechanism map indicating regions involving differing relative contributions of dislocation glide, with zero implying no dislocation-controlled plasticity and unity implying no diffusion-controlled plasticity. The middle contour indicates 50:50 control.

A similar composite model[192] was used to explain the softening with decreasing grain size by assuming that the nanocrystalline material consisted of a grain interior plus an amorphous grain-boundary layer. The grain interior was presumed to deform elastically under stress, while plastic deformation of the grain-boundary layer was governed by Maxwell's equation. Under these assumptions, the strength of the nanocrystalline material decreased linearly with decreasing grain size when that size was below a threshold value. It was pointed out[193] that, in the case of nanocrystalline materials, additional factors need to be considered with regard to grain growth and deformation. For example, grains can increase their size by rotation as well as by curvature-driven motion. This mechanism is especially possible in nanocrystalline material if grain growth is retarded by a finite triple-junction mobility. New deformation mechanisms could then occur, thus leading to the inverse Hall-Petch effect.

The microstructure of a nanocrystalline material can generally be divided into two parts: when the grain size is less than 10nm, the interface can be divided into grain boundaries and triple junctions, thus making the mechanical behaviour of nanocrystalline materials having complicated microstructures very different to that of coarse-grained material. Again considering nanocrystalline material to be a composite, this time involving three phases (grain-core, grain boundaries, triple junctions), a model[194] was proposed for treating the deformation of nanocrystalline material; one which again predicted inverse Hall-Petch behaviour.

It was noted[195] that, when Hall-Petch strengthening predominates and the stress intensity (Hall-Petch slope) is thermally-activated, the strain-rate sensitivity is also predicted to exhibit a Hall-Petch dependence; increasing with decreasing grain size. The activation volume, which is inversely proportional to the strain-rate sensitivity, thus decreases by an order of magnitude. At the smallest grain sizes, the transition to an inverse Hall-Petch dependence has been attributed to the onset of high-temperature grain-boundary weakening. In this situation, the strain-rate sensitivity and corresponding inverse activation volume dependence upon grain size were predicted to exhibit a similar reversal. Another explanation[196] for the inverse Hall-Petch effect has been based upon the statistical absorption of dislocations by grain boundaries, revealing that the yield strength depends upon the strain rate and temperature, and diverges from Hall-Petch behaviour at below a critical grain size. The effect of the grain-size distribution upon the yield stress, ultimate tensile stress and strain of nanocrystalline metals has been shown theoretically[197] to be such that, as the grain-size dispersion increases, the degree of grain-boundary hardening of nanocrystalline materials decreases. The onset of the inverse Hall-Petch effect shifts moreover to smaller grain sizes, and the uniform strain at which necking occurs increases.

A study[198] of the effect of the melting-point upon the Hall-Petch relationship showed that, as the grain-size decreases, the melting-point of nano-structured crystals decreases and the Hall-Petch relationship is no longer satisfactory. When the yield strength or hardness is plotted as a function of the reciprocal of the square-root of the grain-size, it exhibits a maximum, the location of which depends upon the magnitude of the bulk melting-enthalpy of the crystal.

Some studies fail to predict softening in nanocrystalline specimens as grain sizes fall below 10 to 15nm, when they are first thermally relaxed. A simple model illustrated[199] that the increased hardening was the result of grain-boundary relaxation. This suppressed grain-boundary sliding and forced the material to deform via dislocation glide. This explained why some experiments revealed an inverse Hall-Petch relationship, at grain sizes below 10 to 20nm, whereas others did not.

Various disclination-based models have been proposed in order to explain the mechanical properties of nanocrystalline materials. The so-called relay dislocation-disclination model for plastic shear propagation is based upon switching between the translation and rotation deformation mechanisms of plastic deformation. The translation mode is attributed to grain-boundary dislocation sliding, and the rotation mode is attributed to the formation of wedge junction disclinations. These arise from the cooperative processes of intergrain dislocation motion or cooperative grain boundary diffusion. A dependence of the deformation stress upon the grain size, for various grain-aspect ratios, was demonstrated[200] in the case of the relay dislocation-disclination mechanism, and it was shown that the transition from one mechanism to another could contribute to the inverse Hall-Petch relationship.

This relay-race dislocation-disclination model for the development of plastic shear in nanocrystalline materials had been proposed earlier [201] and had been based upon the then-novel mechanism of switching between translational and rotational deformation. The above work was thus a validation of the idea that switching to and fro between the two modes could explain the Hall-Petch effect in these materials.

A modified phase field crystal technique could self-consistently take account of rapid strain-relaxation in parallel with the usual plastic deformation and multiple crystal orientations treated using the usual phase field crystal technique. The present version[202] could be used to study many phase-transformation phenomena which involved rapid strain-relaxation. It was used to study elastic and plastic deformation in nanocrystalline materials, with particular regard to the inverse Hall-Petch effect.

Three different theoretical models were used[203] to treat the inverse Hall-Petch effect at grain sizes below 100nm. These invoked the existence of a high interfacial area and the presence of nanopores. In order to account for interfaces, a gradient plasticity formulation for interface energies was a first step in this direction. A simplified gradient plasticity model which did not contain interfacial terms was combined with a wavelet analysis. Finally, the concept of nano-porosity was used to incorporate this behavior at the nanoscale.

In another approach[204], a viscoplastic constitutive model was based upon a competition between grain-boundary and grain-interior deformation mechanisms; especially the inelastic deformation caused by grain-boundary diffusion, grain-boundary sliding and dislocation movement. The effects of pressure upon grain-boundary diffusion and sliding could also be accounted for. This approach could model the fundamental mechanical behaviour of nanocrystalline metals, including tension-compression asymmetry, enhanced strain-rate sensitivity and both the normal and inverse Hall-Petch effects. A related multi-scale constitutive variational approach[205] modelled nanocrystalline material as a two-phase composite comprising a grain-interior phase and a grain-boundary affected zone which was described in terms of rate-independent isotropic porous plasticity. A crystal-plasticity mechanism then accounted for the transition from partial dislocation-mediated to fully dislocation-mediated plasticity in the grain interior. The constitutive states of both phases were analyzed in the small-strain regime and extrapolated to finite deformation by logarithmic and exponential mapping. By exploiting the rule-of-mixtures, the overall behavior of a grain was deduced by volume-averaging. The extrapolation from a single grain to a polycrystal was achieved via Taylor-type homogenization, assuming a log-normal grain-size distribution. This approach was again able to predict the inverse Hall-Petch effect.

A secant viscosity approach[206], originally formulated for treating the non-linear time-dependent work-hardening creep of dual-phase composites, has been shown to be widely applicable to the modelling of the mechanical properties of a nanocrystalline material. The latter was here conceived as again being a composite involving a stronger grain interior and a softer grain-boundary affected zone. It was thus possible to calculate the grain-size dependence of the flow stress, strain-rate sensitivity and activation volume, and to be used to explain why the flow stress first increased, and then decreased, as the grain size changed from coarse to nanometric … leading to the Hall-Petch and inverse Hall-Petch effects. The critical point, at which the slope of the strength variation with respect to grain size became zero, was found to correspond to the strongest material state. The strain-rate sensitivity exhibited a similar behaviour to that of the flow stress, but the activation volume exhibited an exactly contrary behaviour.

A model was proposed[207] for predicting the critical grain-size at which nanocrystalline face-centred cubic metals attain a maximum steady-state flow stress. As usual, the model

Materials Research Forum LLC
https://doi.org/10.21741/9781644900352

considered the nanocrystalline metal to comprise two so-called phases; one being the grain-boundary and the other the grain interior. The former phase was assumed to possess deformation characteristics which were different to those of the latter phase. The critical grain-size which corresponded to the maximum steady-state flow stress was predicted to decrease with deformation temperature, and to increase with strain-rate. Both the normal and inverse Hall-Petch relationships could be described by this model.

A gradient plasticity model was proposed[208] which could reproduce the anomalous softening at minute grain-sizes by treating the grain boundary as a separate phase of finite thickness. The theoretical expression which was deduced for the yield stress as a function of grain-size could explain the experimental data which exhibited the normal-to-abnormal Hall-Petch transition, and predict the grain size at which the transition occurred. Analytical expressions which were derived for the flow stress in nanomaterials were in precise agreement with atomistic simulations which predicted that, at below a critical grain-size, the flow stress decreased proportionately.

The true ultimate tensile strengths of 11 metals, in amorphous and nanocrystalline states resulting from severe plastic deformation, were calculated[209] for 0 and 298K. The results for amorphous samples could be correlated with the yield strengths of nanocrystalline samples, and a reason for deviations from the Hall-Petch relationship was advanced which was supported by calculations of the diffusion characteristics.

In recent years, the effect of temperature upon the inverse Hall-Petch effect has been investigated by using the phase-field simulation technique, with the results indicating that the effect is weaker at low temperatures. The results also indicate[210] that a change in the microscopic deformation mechanism as a function of temperature is the main reason for the weakening. At high temperatures, grain-rotation and grain-boundary migration markedly reduce the yield stress, such that a nanocrystalline material exhibits the inverse Hall-Petch effect but, at low temperatures, both grain-rotation and grain-boundary migration occur with great difficulty and the dislocations which are nucleated from the cusps of serrated grain boundaries instead become active. The absence of grain rotation and of grain-boundary migration during deformation is largely responsible for the weakening of the inverse Hall-Petch effect. Because a small grain-size is favourable to grain-boundary migration, the amount of weakening decreases with decreasing average grain-size at low temperatures. Assuming that the inverse Hall-Petch effect can be attributed to mesoscopic grain-boundary sliding-controlled flow at the scale of the grain-diameter, equations can be derived for estimating the free-energy of activation of the rate-controlling process, for estimating the free volume fraction involved in sliding and for estimating the average number of grain boundaries which align to form a planar interface during superplastic deformation. The agreement between prediction and experimental data is found to be satisfactory[211]. The scope of molecular dynamics methods has expanded from treating a few atoms to handling grain sizes of up to 50nm at strain-rates of more than 10^6/s. In the latter regard, the occurrence of twinning at the high strain-rates occurring during shock compression has been considered[212]. Similar modelling has shown that the inverse Hall-Petch behavior can be attributed to the relief of stress build-up at grain-boundary junctions by the emission of dislocations. It is noted[213] however that pure grain-boundary sliding, with no further plastic accommodation by dislocation emission, is grain-size independent.

Each grain of a nanomaterial exhibits features such as faces or triple-junctions and, during grain growth, the number of such features per grain and the cumulative features of all of the grains, change randomly with time. Models existed which described changes in the number of features per grain during grain-boundary migration and grain rotation-coalescence, but did not treat changes in the cumulative features. Stochastic theory and the random marked point field technique were therefore used[214] to treat the temporal and thermal evolution of random cumulative features on grains in nanomaterials. It was found that the mean number of features per grain increased, while the density of grains decreased, during grain growth. An increase in annealing temperature resulted in a relatively higher increase in the mean number of features per grain, in a further decrease in grain density, in a relative increase in mean cumulative features and in variable dispersions of cumulative features. The evolution of the statistics of cumulative features depended upon the nature of a diffusion term, on the critical number of faces per grain and on the type of grain-growth mechanisms. For some choices of the diffusion term, the dispersion of cumulative features changed in a similar manner to that of the mechanical properties as given by the normal and inverse Hall-Petch relationships.

Nanocrystalline material having an average grain size ranging from 11.61 to 31.32nm was modelled[215] using the phase field crystal model. The simulation results showed that grain-rotation and grain-boundary migration were mainly responsible for microscopic deformation. A small grain-size was favorable to grain rotation, and this could reduce the yield strength such that the material exhibited an inverse Hall-Petch effect. When the grain size was very small, dislocation activity began to occur. Due mainly to changes in the grain-boundary structure, the latter could make a finite contribution to deformation. With increasing grain size, grain rotation became difficult, and grain serration and dislocation-emission occurred.

Finite-element modelling[216] of the effect of grain-size uniformity on the flow stress of nanocrystalline metals again provided clear evidence of an inverse Hall-Petch relationship which was attributed to the effect of macroscopic strain accumulating preferentially at grain boundaries. The important effect of the grain-size distribution upon the flow stress of polycrystals was again demonstrated.

A continuum elasto-viscoplastic fast-Fourier-transform model has been used[217] to study the role played by the gradient of rotation upon the local mesoscale and macro-scale mechanical responses of nanocrystalline material. This led to the derivation of an extended periodic Lippmann-Schwinger equation which accounted for couple stress equilibrium such that, in addition to the usual boundary conditions on strain-rate and Cauchy stresses, the model permitted the imposition of additional non-standard couple stress and gradient of rotation boundary conditions. This showed that elastic and plastic curvatures accommodated some of the local and macroscopic Cauchy stresses. The grain-boundary interfaces could also be characterized by using curvatures which reflected their structure and defect content. Depending upon the magnitude and distribution of said curvatures, local stresses which were generated in the grain-boundary neighborhood then activated slip systems which were in addition to those satisfying the Schmid criterion. It has been suggested that the generation of both polar dislocations and disclinations contributes to the plasticity of nanocrystalline materials. At the macro-scale, this could then result in a strain-rate dependent weakening; again explaining the inverse Hall-Petch effect.

In a further twist, experiments performed on ultrafine-grained metals reveal an abnormal mechanical response to intense dynamic loading. On the basis of conventional plasticity theory, which takes account of dislocation slip and grain-boundary sliding, it is shown that such a response is normal in these structures. Calculations predict an inverse Hall-Petch effect to occur in ultrafine-grained metals at strain-rates of above 10^7/s, whereas conventional low strain-rate experiments and molecular dynamics simulations find such an inverse Hall-Petch relationship only in nanocrystalline materials. It is concluded[218] that ultrafine (100-200nm) grained metals have a maximum dynamic shear strength.

Local models cannot however take account of any non-uniformities, and can properly describe behavior only in the case of homogeneity. The effect of heterogeneity can be incorporated by supplementing phase-mixture models using gradient terms of second and fourth order[219]. As usual, differing deformation mechanisms are assumed to operate concurrently within the grain interior and at the grain boundaries. Deformation in the grain boundaries is attributed to diffusional mass transport while, within the grain, dislocation-glide and diffusion are occurring. As shown above, such a model is well capable of correctly predicting the transition from Hall-Petch behavior in the normal grain-size range to inverse Hall-Petch behaviour in the nanocrystalline grain-size range. The inclusion of second-order and fourth-order strain gradients now permits non-uniformities within the strain field to represent strain patterns, in combination with a regularizing effect.

The Ubiquity of the Inverse Hall-Petch Law

Just as the data on the normal Hall-Petch law were reviewed, it will be useful to survey its prevalence, according to either experimental data or theory, on the same materials; this time beginning with the face-centred cubic metals.

Gold

Molecular dynamics simulations have been used[220] to study the effects of grain-size and ligament-diameter upon the mechanical properties of nanocrystalline nonporous gold. These revealed that the main deformation mechanism was a combination of grain-boundary sliding, grain rotation and dislocation movement. Uniaxial tensile tests indicated the occurrence of an inverse Hall-Petch relationship between the strength and nominal grain-sizes of 7.9 to 52.7nm. An increase in flow stress was tentatively attributed to a lower total fraction of grain-boundary sliding and grain rotation during the deformation of samples having larger grain sizes. The Young's modulus was linearly proportional to the inverse grain-size. That depended largely upon the volume fraction of grain boundaries, and thus a decrease in grain-size led to a relatively lower Young's modulus. Simulations of samples having ligament diameters ranging from 4.07 to 8.10nm showed that an increased ligament diameter resulted in a decreased flow stress and in an increased Young's modulus. In another molecular dynamics simulation[221], incipient plasticity and void nucleation in nanocrystalline gold was investigated. The stress-strain curve at 300K showed that the maximum stress corresponded to a grain-size of 3.2nm. This was much lower, and the stress curve very different, when compared with those corresponding to other grain sizes. The Young's modulus increased with increasing mean grain-size; that is, an inverse Hall–Petch behaviour occurred. Slip was the main deformation mode at a mean grain-size of 3.2nm. The internal stress became more pronounced with increasing temperature. At 700K, the main deformation zone was concentrated at the middle of the samples, resulting there in an almost force–induced structural transformation. During tensile testing, void damage occurred at the junctions of three grain boundaries. With decreasing mean grain-size, less internal differential slip was generated at a given temperature and strain.

The effect of alloying has naturally been explored. The mechanical behavior of electrodeposited nanocrystalline alloys containing up to 20wt% of copper was assessed[222] as a function of strain-rate sensitivity by using foil samples having grain sizes which were as small as 3nm. Micro-scratch measurements showed that the hardness approached the ideal value as the grain size decreased to 7nm. At smaller grain sizes there was a decrease in strength and thus a deviation from the Hall-Petch law. This was attributed to an increase in the activation volume for deformation as the grain size was further decreased.

Silver

The tensile deformation of polycrystalline silver nanowires having various grain-sizes was modelled[223] using molecular dynamics simulation. The model predicted that the wires would exhibit softening at grain-sizes which were less than 13.49nm, and thus an inverse-Hall-Petch behaviour. Plastic deformation at that point was dominated by sliding at the grain boundaries and by rotation of the grains. Five-fold twins were formed in a later deformation stage. The plastic deformation mechanism changed to dislocation sliding when the grain-size was greater than 13.49nm, and large numbers of twins formed in the later stages of deformation.

Palladium

In an early study[224], room-temperature tensile and creep tests were performed on samples of nanocrystalline palladium. For comparison purposes, tensile tests were also performed on coarse-grained material. The latter measurements were affected by the large size of the grains with respect to the sample dimensions, while the nanocrystalline samples yielded bulk properties. The strain-rate in the tensile tests was about 2×10^{-5}/s. The results showed that nanocrystalline specimens, with a grain size of 7 to 10nm, were very strong. The strength advantage over the coarse-grained material was attributed to the ultra-fine grain-size, but the data were insufficient to test the validity of the Hall-Petch relationship down to nanocrystalline grain-sizes. A model was proposed[225] in order to explain the behaviour of nanocrystalline materials, with the distribution of grain sizes being simulated by a logarithmic normal distribution, and assuming one dislocation per grain. Plastic yielding was taken to be controlled by the stress required to produce dislocation loops in a set of larger grains. Dislocations in the remainder of the smaller grains were considered to be in a sub-critical configuration, thus producing reversible deformation and contributing only to inelastic deformation. The model's predictions were in very good agreement with the yield-stress versus grain-size relationship for 5 nanocrystalline materials. Among these, 3 materials exhibited an inverse Hall-Petch slope while the others had a positive Hall-Petch slope. The model also predicted a decrease in Young's modulus with decreasing grain size. Plastic deformation of Pd-10at%Au alloy with an average grain-size of 14nm was investigated[226] in compression at temperatures ranging from 4.2 to 300K. Decreasing the grain size from 10μm to 14nm resulted in a 4.7 to 6.4 increase in the applied stress as the temperature was decreased from 300 to 77K. A further decrease in temperature did not lead to any additional increase in the applied stress. The samples exhibited an extended micro-plasticity stage, with parabolic strain-hardening at up to 4.2% strain at room temperature. With decreasing temperature, the strain-range over which micro-plasticity occurred became smaller and was only 2% by 40K. When deformed in the macro-plastic range, the specimens exhibited weak strain-hardening at room temperature and at 210K, while strain-softening occurred at temperatures down to 40K. The strain-hardening was associated with marked grain growth, indicating an inverse Hall-Petch relationship. The deformation curves exhibited serrated plastic flow at 10K. A deformation mechanism was proposed in which the plasticity was governed by grain-boundary sliding, and accommodated by the slip of dislocations emitted from grain boundaries: grain-boundary sliding permitted the stress concentration required for dislocation emission, while dislocations offered a means for accommodating any geometrical incompatibility which might arise along the grain-boundary sliding path.

Specimens of $Pd_{80}Ni_{20}$ were irradiated[227] with 70, 80, 90 or 100 laser pulses (10ns, 50mJ, 10Hz) in air using a Q-switched 1064nm Nd-YAG laser. The specimen surface exhibited wave-like structures, ripples, dips, ridges, microcones, flakes and fissures. The peak intensity of the (111), (200), (220), (311) and (222) planes increased linearly with the number of laser pulses, showing that the concentration of point defects such as vacancies progressively decreased with increasing number of pulses. This was attributed to annealing. The average surface hardness first decreased after irradiation with 70 pulses, but later increased linearly with the number of pulses, up to 100. An inverse-Hall-Petch relationship was obeyed for crystallite sizes ranging from 19 to 27nm. As the crystallite size decreased from 27 to 19nm, the volume fraction of amorphous phase gradually increased.

Aluminium

The behavior of a three-dimensional nanocrystalline material under uni-axial tensile loading was modelled[228] by means of molecular dynamics simulations involving 48 aluminium grains and 551011 atoms. The elastic moduli were intermediate between those of the Voigt and Reuss models. Crystal slip, caused by dislocations emitted from grain boundaries, was one of the main deformation mechanisms. It was concluded that crystal slip and grain rotation occurred during atomistic deformation and played an important role in the inelastic deformation of nanocrystalline material, including deviations from Hall-Petch behaviour. In similar later molecular dynamics simulations[229], account was taken of the stacking-fault energy using three different models. In each case, the dependence of the maximum stress upon grain size was described by an inverse Hall-Petch relationship. This tendency could be explained directly in terms of the volume-effect of grain boundaries. Crystal slip and grain-boundary sliding occurred, but grain-boundary sliding predominated in a small-grain model. Most of the crystal slipping was caused by the motion of perfect dislocation when one sort of potential was assumed but, when a different potential was used, most of the slipping was caused by the motion of Shockley partial dislocations. Because the latter potential underestimated the stacking-fault energy, the core-length of the extended dislocation was comparable to the grain-size and many of the stacking faults therefore remained within grains. During deformation via partial-dislocation motion, a strange grain-rotation occurred. The deformation of nanocrystalline materials was concluded to be strongly influenced by the stacking-fault energy. Grain-switching was also predicted, similar to the Ashby-Verrall mechanism. The latter mechanism was however based upon diffusion, whereas grain-boundary migration and sliding predominated in the present case. Other molecular dynamics simulations revealed[230] that a strain-rate dependence was mirrored by the critical resolved shear stress for dislocation propagation. The strain-rate and temperature-dependence, together with stress and grain-size dependences, supported the assumed importance of thermally-activated dislocation-mediated plasticity in the inverse Hall-Petch regime. Modelling demonstrated that the maximum material strength could be explained on the basis of dislocation mechanisms alone. Molecular-dynamics simulations of tensile-loaded nanocrystalline aluminium were performed[231] using an embedded-atom-method potential. The use of two different sample-preparation methods permitted the comparison of mechanical properties as a function of sample quality. A Voronoi-based polycrystal contained essentially no pores and exhibited different mechanical properties to those of material which was sintered from spherical nanoparticles; resulting in a lower-density material. An inverse Hall-Petch behaviour of the flow stress was again found for grain sizes smaller than 10nm.

Molecular dynamics simulations[232] of the surface roughening of nanocrystalline samples showed that the roughness increased with grain-size and strain. Differing strain distributions in the grains and grain boundaries at the sample surface were noted, leading to the formation of local roughness. A linear relationship existed between the degree of roughness and the out-of-plane strain component. An elastic-plastic transition was detected at about 3.5% strain, and an inverse Hall-Petch effect was observed.

Molecular dynamics simulation was used to explore[233] the effect of the twin-boundary spacing upon crack propagation in aluminium. This revealed that the orientation of the initial crack could easily influence the crack-growth mechanism. Twin-boundaries strengthened the aluminium, when the crack was oriented parallel to the twin boundary, because the twin boundary hindered the emission of dislocations. There was moreover an optimum twin-boundary spacing, demarcating inverse- and normal Hall-Petch effects, when the twin-boundary spacing was close to a critical twin-boundary spacing. There was also a yield-strength hardening effect in nanotwinned aluminium when the crack was oriented perpendicular to the twin boundary, and the Young's modulus was smaller for nanotwinned aluminium than for twin-free aluminium. This distinctive behavior was attributed to dislocation nucleation and the repulsive effect of twin-boundaries on dislocations and crack propagation, as well as to the distance between a crack-tip and a twin boundary.

The formation of solidification defects, and their behaviour during the uniaxial tensile deformation of solidified polycrystalline aluminium, were modelled[234] by means of molecular dynamics simulation. Solidification was simulated for both isothermal and quenching conditions. In the initial stages of nucleation, coherent twin boundaries and five-fold twins formed; depending upon the quenching-rate or undercooling, and the solidified material contained randomly distributed grains, twin boundaries and vacancies. Void formation at grain boundaries, and de-twinning of pre-existing and deformation twins, occurred during uniaxial deformation. The deformation temperature exerted a greater effect, than did the applied strain-rate, upon the strength of solidified samples. When the grain-size was smaller than 10nm, the yield strength and Young's modulus increased with increasing grain size, according to the inverse-Hall-Petch relationship. The molecular dynamics simulations also predicted a higher yield strength for monocrystalline aluminium, and a high plastic deformation for polycrystalline aluminium.

Bulk cryomilled aluminium, stabilized by diamantane nanoparticles, was compared[235] with bulk cryomilled commercial-purity aluminium which contained no diamantane. In each case, the cryomilled aluminium was consolidated by hot isostatic pressing followed by high-strain extrusion at room temperature. Whereas the grain-size (155nm) of cryomilled commercial-purity aluminium was larger than that (68 to 95nm) of cryomilled aluminium containing diamantane, the strength of the former was greater. This was explained in terms of the occurrence of nanoscale softening; that is, a transition from the usual Hall-Petch behavior to the inverse Hall-Petch behavior. In another cryomilling study, hot isotropic pressing and hot extrusion were used[236] to prepare bulk nanocrystalline material, and the contributions made to the yield strength by grain-refinement, Orowan strengthening and dislocation strengthening were estimated quantitatively. The average grain-size was about 300nm, and the density and microhardness were 2.692g/cm^3 and 109.15H$_V$, respectively. The sample had a yield strength of 270MPa and an ultimate strength of 379MPa, with an elongation of 3.2%. The high strength was attributed mainly to the grain-size effect and to Orowan strengthening.

It has been remarked[237] that models which relate grain-size to mechanical properties tend to consider only the equivalent radius and ignore other dimensions such as, for example, the semi-minor axis, the semi-major axis and other radii which define the grain shape. It was noted that grains elongate up to a maximum value and then decrease, with further grain refinement, due to grain breakages. The yield stress increases with elongation up to a maximum and then continuously decreases. The various approaches to measuring grain radius exhibit a common trend with regard to the normal Hall-Petch and inverse Hall-Petch relationships, but with differing critical grain-sizes. Materials comprising high-curvature grains have a higher yield stress. Reducing the strain-rate leads to a higher yield stress, with a critical strain-rate beyond which further reductions do not lead to stress enhancement.

Molecular dynamics simulation was used[238] to model the uniaxial tensile deformation of nanocrystalline samples having a wide range of grain-sizes. The largest sample space used had a mean grain size of about 30nm and contained more than 100 million atoms. Deformation-twinning and grain-boundary migration were the main mechanisms of tensile deformation. A complete Hall-Petch relationship was derived which comprised 3 distinct regions: normal, inverse and extended. In a molecular dynamics simulation[239] of rapidly solidified aluminium containing multiply-twinned nanograins, mean grain-sizes ranging from 3.1 to 24.4nm were assumed to be present. The properties of rapidly solidified aluminium containing multiply-twinned nanograins depended upon the grain size. The flow stress obeyed an inverse Hall-Petch relationship within the above grain-size range. Specimens having large and small grains exhibited differing deformation mechanisms during uniaxial tensile deformation. At large grain sizes, the predominant deformation mechanism was twin-boundary migration, but co-existed with dislocation activity. The morphologies of twinned nanograins remained almost unchanged during deformation. At small grain sizes, deformation was governed by the softening effects of grain growth and de-twinning. Grain-boundary migration and grain-rotation occurred during grain growth. The multiply-twinned nanograins gradually transformed into parallel twins or monocrystal.

The behaviour of indented aluminium/titanium multi-layer films has been monitored[240] by means of Auger electron spectrometry, focused ion-beam machining and transmission electron microscopy. Pure and multi-layered films, on Si(100) substrates, were prepared via radio-frequency magnetron sputtering. The normal Hall-Petch relationship and its inverse were seen in the nano-indentation load-displacement curves, hardness and Young's moduli data for individual layers with thicknesses of 28, 14 or 7nm.

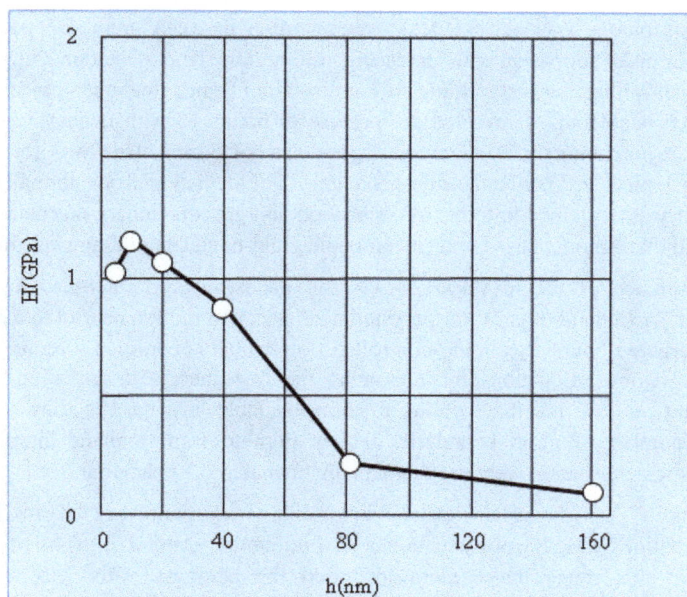

Figure 13. Relationship between the hardness of multilayered
Al-3.3at%Zr films as a function of columnar crystal height

Multilayered films having Al-Zr crystalline layers of various thickness, plus amorphous layers of identical thickness, were studied[241] in which the heights of columnar crystals in the crystalline layers ranged from 5 to 160nm while the diameter was maintained at 10 to 15nm; regardless of the height. This permitted control of the grain size independently of other microstructural parameters. Analysis of the mechanical properties showed that the inverse Hall-Petch relationship also occurred in Al-Zr nanocrystalline alloys (figure 13). The critical grain sizes for deviation from the Hall-Petch relationship and towards the inverse Hall-Petch law were about 40 and 10nm, respectively.

Powder samples of as-cast Al-25at%Fe and Al-34.5at%Fe, close to the Al_3Fe and Al_2Fe intermetallic phases, were subjected[242] to high-energy ball-milling for up to 50h, resulting in a transformation from monoclinic (Al_3Fe) and triclinic (Al_2Fe) structures to orthorhombic (Al_5Fe_2), and from crystalline to amorphous. Hardness measurements revealed a change from hardening to softening during prolonged milling. The softening effect in milled powders having a composite structure involving nanocrystalline and amorphous phases was attributed to a competition between grain-size reduction and amorphous phase formation with increasing milling time. An indentation study[243] of high-energy ball-milled nanocrystalline Al_5Fe_2 itself confirmed the occurrence of inverse Hall–Petch behaviour. It revealed an increase in hardness, with decreasing grain size; reaching a maximum of 9.0GPa for a grain-size of 32nm. This was followed by a decrease; typical inverse-Hall-Petch behaviour. The deviation from normal Hall–Petch behaviour was explained in terms of dislocation and grain-boundary mechanisms. It was concluded that a model based upon mesoscopic grain-boundary sliding was most likely.

The occurrence of the inverse-Hall-Petch in nanostructured intermetallics has been attributed[244] to a hand-over of intra-crystalline dislocation motion control to a mesoscopic grain/interphase boundary sliding-controlled flow which is confined to grain/interphase-boundary regions. Equations for estimating the free energy of activation of the rate controlling process, the free volume fraction available in a basic sliding unit and the average number of grain boundaries which align to form a plane interface during superplastic deformation were used to explain the material's behaviour.

The validity of the Hall-Petch relationship was confirmed[245], in ball-milled nanocrystalline AlMg4.8 powder, down to a minimum grain size of 44nm. Prolonged milling in this range nevertheless increased the hardness. This was attributed to contamination effects and to the effect of perfect and partial dislocations. Recovery occurred at 100 to 230C, and appreciable grain growth began at 250C. The enthalpies of recovery and grain growth were 39 and 208J/mol, respectively. Dynamic strain-aging was described by an activation energy for recovery of about 120kJ/mol. When electro-discharge consolidation was used[246] to prepare nanocrystalline powders the products, because of the markedly small resultant grain size, exhibited an inverse Hall-Petch behaviour.

Mechanical alloying of $Al_{100-x}Fe_x$, where x was 2.5, 5, 10, 15 or 20, was carried out for 20h, followed by cold-consolidation[247]. This led to the formation of supersaturated solid solutions of iron in aluminium, for 2.5 to 10at%Fe alloys, while the Al-20at%Fe alloy formed nanocrystalline metastable Al_5Fe_2. During mechanical alloying, an intermetallic phase formed only when the crystallite size fell below 20nm. The final crystal size following 20h of mechanical alloying was related to the iron content. The Hall-Petch slope decreased below a crystallite size of about 15nm (figure 14), suggesting a possible change from hardening to softening. A microhardness of 12.4GPa was found for nanocrystalline Al-20at%Fe alloy following annealing (673K, 2h).

Materials Research Forum LLC
https://doi.org/10.21741/9781644900352

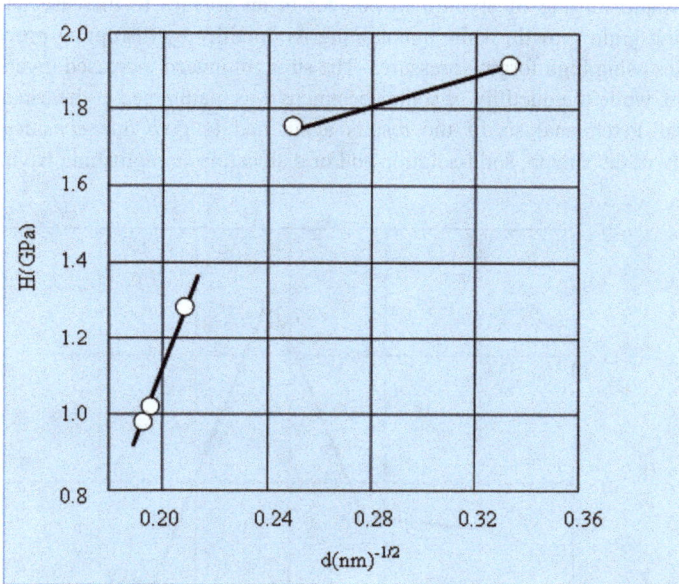

Figure 14. Hardness of Al–Fe alloys as a function of grain-size

Bulk nanocrystalline Al-4Cu alloy, having a grain-size ranging from 47 to 105nm, was prepared[248] by mechanical alloying followed by vacuum hot-pressing at various temperatures. The Hall-Petch plot indicated a frictional stress of 170MPa and a positive slope of 0.13MPa√m. Pure aluminium, for comparison, had a frictional stress of 15 to 30MPa and a slope of 0.06 to 0.09MPa√m. The high values of frictional stress and slope, in the case of the alloy, were attributed to the presence of Al_2Cu precipitates and oxide particles. As a more extreme example of the effect of particles, metal-matrix composites were prepared[249] by first cryomilling inert-gas-atomized AA5083 powders with B_4C particles, thus producing agglomerates of nanocrystalline aluminium grains containing uniformly dispersed and bonded sub-micron B_4C particles. These agglomerates were then blended with coarse-grained aluminium powder and consolidated to form a bulk composite. These specimens possessed a high strength which was attributed to the combined effects of Hall-Petch strengthening arising from nanocrystalline aluminium grains, of Zener pinning by B_4C particles, of interfaces between the three constituents and of a high dislocation density. Ultra fine-grained AA5083 plate was prepared[250] by the quasi-isostatic forging of cryomilled powder in an attempt to increase its strength by minimizing grain growth, while maintaining its ductility by disrupting previous particle boundaries using high forging pressures. The strength indeed increased inversely with the grain-size, while the ductility of some specimens was maintained at the usual level of the alloy. Hall-Petch analysis of the results again had to give due consideration to the possibility of dispersion, solid-solution and/or dislocation strengthening having occurred.

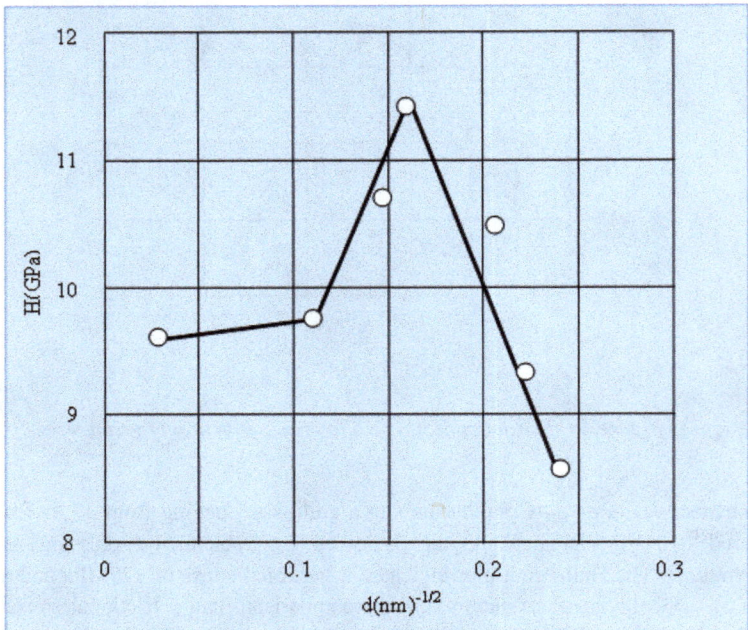

Figure 15. Hall-Petch plot of the hardness of $Al_{62.5}Cu_{25}Fe_{12.5}$ quasicrystal

How might the Hall-Petch relationship operate in quasicrystalline material. An inverse Hall-Petch behavior has been reported[251] to occur in nano-quasicrystalline $Al_{62.5}Cu_{25}Fe_{12.5}$. Powders having various grain-sizes were produced by the mechanical milling of spray-formed quasicrystals. The hardness increased with decreasing grain-size down to about 40nm and then decreased with further refinement (figure 15), thus revealing inverse-Hall-Petch behaviour. This critical grain-size was larger than that for other metallic nanocrystalline alloys. The inverse behaviour was attributed to the

Materials Research Forum LLC
https://doi.org/10.21741/9781644900352

structural complexity of quasicrystals and to thermally-activated shearing of atoms at the grain boundaries.

A generalized theory of steady-state deformation has been based[252] upon experimental data for dilute single-phase aluminium alloys. It was hypothesized that, although properties such as flow-stress and grain-size may remain constant in time, there can be a continuous loss of grain boundaries in the steady state. A proposed model took account of the activity of grain-boundary dislocations in order to describe the kinetics of steady-state deformation. The steady-state, as a function of strain-rate and temperature, then defined the limits of conventional grain-size and strength relationships. That is, the Hall-Petch law holds when the grain-size is greater than that in the steady state, while the inverse-Hall-Petch relationship replaces it if the grain-size becomes smaller than the steady-state value. It is concluded moreover that the transition between the two relationships depends upon the deformation conditions rather than being an intrinsic material property. A general deformation scaling law was defined. It was argued to be unconvincing to describe a behavioural change as being a function of scale, by using that very same scale as a criterion. It was noted that strength versus grain-size dependences, like the inverse Hall-Petch law, have been observed under creep and superplastic flow conditions where the grain-size ranged from nanometres to millimetres, but without there being any clear underpinning in those cases. It tends to be assumed that the dynamic response of a microstructure to plastic deformation is strongly dependent upon the grain-size, but grain sub-division can occur during the deformation of coarse-grained materials, while starting with small grains may engender grain-coarsening. The scaling law defined here suggested that the normal-inverse changeover in the steady state is governed by the instantaneous grain-size, as it were, rather than by any fixed grain-size. Thus, for a given initial grain-size, hardening occurs if the steady-state size defined by the imposed strain-rate and temperature is smaller; and softening occurs in the opposite case.

Copper

Computer simulations were made[253] of the deformation of nanocrystalline copper, showing that the main deformation mode was sliding in the grain boundaries via a large number of uncorrelated events in which a few atoms slid relative to each other. Little dislocation activity occurred in the grain interiors. Localization of deformation to the grain boundaries led to hardening as the grain-size was increased. This inverse Hall-Petch effect implied that a maximum in hardness existed for some grain size. The softening seen at small grain sizes was attributed to the larger fraction of atoms at grain boundaries. With increasing temperature, the material became softer in both the plastic and elastic regimes, and porosity also led to softening of the material. It was noted prophetically[254] that the softening which occurred at very small grain-sizes would ultimately impose a limit on how strong nanocrystalline metals could be made.

Molecular dynamics simulations were also made[255] of polycrystalline copper having a bimodal structure. Using a grain-size distribution based upon the inverse-Hall-Petch law, small-grain and large-grain regions were found to play differing roles in uniaxial tension. The large-grain region imparts a high stress for lattice-dislocation activation.

Molecular dynamics simulations of uniaxial tensile testing showed[256] that the elastic modulus gradually increased with increasing mean grain-size, that the flow-stress increased in parallel and that the flow-stress was proportional to the square-root of the grain size; indicating an inverse-Hall-Petch relationship. Meanwhile the elastic modulus decreased with increasing ambient temperature and the flow-stress was negatively correlated with the temperature. The simulations confirmed that the mechanism of plastic deformation in polycrystalline copper having a grain-size of 4.65 to 9.31nm was predominantly grain rotation and grain-boundary sliding, with dislocation nucleation and migration no longer being the dominant factors in plastic deformation.

When samples with a density greater than 98.5% and grain sizes of 20 to 50nm, were prepared[257] by gas-condensation and *in situ* hot-compaction, the Vickers microhardness of samples compacted using pressures of 1.0 to 3.5GPa at 423K was greater than that of coarse-grained (50μm) copper. No appreciable grain growth was observed. The hardness appeared to increase with increasing grain size between 20 and 24.8nm, in contrast to the normal Hall-Petch law which was obeyed between 20 and 50nm. This was attributed to a higher density, smaller grain size and the effect of internal micro-strain.

The grain-size of copper thin films was controlled[258] by introducing a few monolayers of an insoluble element, such as tungsten, having a high surface energy. This increased the interfacial energy and provided more nucleation sites. The grain size was governed by the copper-layer thickness and by the substrate temperature at which it transformed into islands of almost uniform size. In this way, the grain-size could be reduced from 160nm to 70 to 80nm and even to 4nm. An appreciable increase in hardness occurred, from 2.0GPa for 180nm films to 12.5GPa for 7nm films, but there was a decrease in hardness below 7nm. The increase in hardness with decreasing grain size could be explained as usual in terms of the Hall-Petch model. The decrease in slope, and eventual decrease in hardness below a certain grain-size, had to be explained by using a new model which was based upon grain-boundary sliding.

The uniaxial tensile deformation of nanocrystalline copper was modelled[259] by using molecular dynamics simulations with a Finnis-Sinclair potential, and the mean grain-size of the simulated material was varied from 5.38 to 1.79nm. The strength and Young's

Materials Research Forum LLC
https://doi.org/10.21741/9781644900352

modulus were strongly dependent upon the grain size and the nanocrystalline structure, and exhibited an inverse Hall-Petch effect.

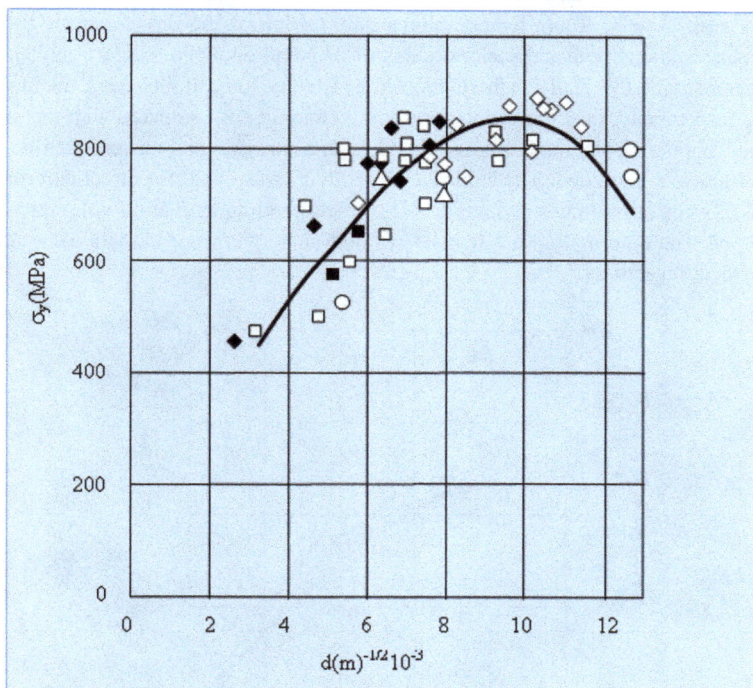

Figure 16. Comparison of experimental data with inverse-Hall-Petch model. Heavy line: present model, open squares: Sanders (1997), open diamonds (Fougere (1992), open circles: Nieman (1991), triangles: Chokshi (1989), filled diamonds: Weertman (1993), filled squares: Chen (2006)

The Vickers hardness of electrodeposited copper films[260] depended upon the processing conditions. Nanocrystalline film with a grain size of 31nm had a hardness which deviated from the Hall-Petch relationship because the grain-size dependence of the hardness was smaller for grain sizes below 100nm than for those greater than 100nm. The Hall-Petch constants for films which had been processed using thiourea were also different to those for films which had been processed using gelatine. These differences were not attributed to the texture but rather to the presence of super-abundant vacancies which had been generated during electrodeposition. Two types of nanocrystalline copper were prepared[261] by direct-current or pulsed-current electrodeposition, giving mean grain-sizes of 65 and 33nm, respectively. Room-temperature tensile testing of the direct-current and pulsed-current specimens revealed yield stresses of 332 and 545MPa, respectively; in general agreement with the Hall-Petch relationship. The fracture strains were about 18.2 and 3.4%, respectively; thus indicating very limited ductility as compared with that of coarse-grained copper. Both materials exhibited higher stresses and lower ductilities as the strain-rate was increased. The strain-rate sensitivity was 0.020 for direct-current samples and 0.029 for pulsed-current samples. The corresponding activation volumes were $90b^3$ and $36b^3$. The deformation mechanisms of both materials were thought to be dominated by dislocation activity.

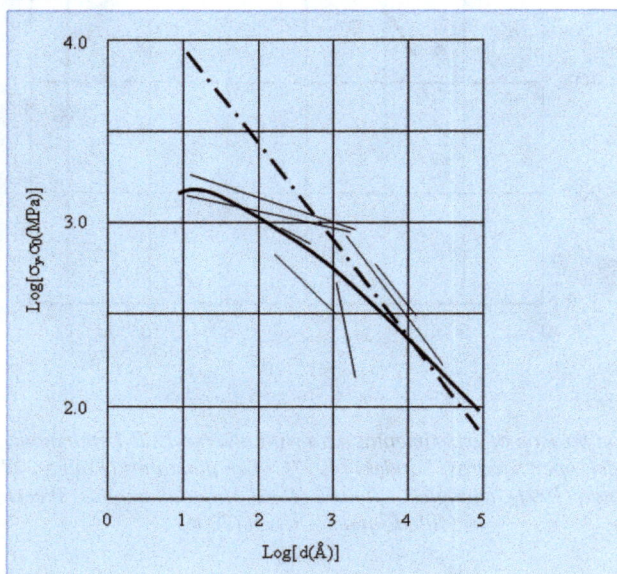

Figure 17. Dependence of the yield stress on the microstructural length-scale. Dash-dot line: extrapolated Hall-Petch law. Heavy solid line: best fit to previous data. Other lines: previous data.

A model for the inverse-Hall-Petch effect in nanocrystalline material was proposed[262] which assumed that lattice distortion along grain boundaries could produce internal stresses and promote grain-boundary yielding. The model was applied to nanocrystalline copper data (figure 16), and the minimum grain-size at which the inverse- Hall-Petch effect was expected to appear was here estimated to be about 11nm.

The yield strength and hardness of copper-based materials exhibited[263] a progressive decrease in strength, with respect to the extrapolated Hall-Petch relationship, as the length-scale of the microstructure decreased. The overall trend could be modelled by scaling the elastic screening length, for dislocation line tension, with the microstructural length-scale, according to:

$$\sigma_y - \sigma_0 = kd^{-1/2}\ln[d/r_{eff}]$$

where r_{eff} was the dislocation cut-off radius of Scattergood and Koch[264]. When k was taken to be 2.7 x 10^4Pa$\sqrt{m}$, σ_0 was 168MPa and r_{eff} was 1.7Å, a large number of previous data could be used to show the deviation from the Hall-Petch law (figure 17).

Atomic-scale simulations were made[265] of nanocrystalline copper having grain-sizes ranging from 5 to 50nm. These revealed that a clear maximum in the flow stress occurred when the grains were 10 to 15nm in diameter, and there was also a change in the deformation mechanism: from dislocation-mediated plasticity at larger grain-sizes to grain-boundary sliding at smaller grain-sizes. At above the maximum hardness, the grain-size dependence of the hardness was consistent with the Hall-Petch relationship and the concept of dislocation pile-ups in grains. It had not been clear whether this explanation of the effect was valid for sub-micron grains, but these simulations clearly showed the existence of pile-ups in simulations for an average grain-size of 50nm. Dislocation motion in the grains was dominated by the grain boundaries because essentially all dislocation nucleation and absorption occurred at those boundaries. During plastic deformation, large numbers of stacking-faults and a much smaller number of twin boundaries, were created. These did not contribute greatly to the flow stress, while the number of stacking-faults increased with strain. In order to model the occurrence of the inverse-Hall-Petch law in nanocrystalline material, the latter was treated[266] as being a composite made up of grain cores and grain boundaries. Deformation of the former involved a viscoplastic component which took account of the dislocation glide mechanism and Coble creep. The boundary was treated as an amorphous material which exhibited perfect elastic-plastic behavior. A very similar two-phase composite scheme was used[267] to model the transition from Hall-Petch to inverse-Hall-Petch behaviour and relate it to a change in plastic deformation mode in one phase, from a dislocation-glide driven mechanism to a diffusion-controlled process.

A suggested constitutive law for quantifying the macroscopic effect of grain-boundary dislocation emission upon the behavior of pure face-centered cubic nanocrystalline materials proposed[268,269] that an emitted dislocation ended its trajectory in the grain boundary opposite to the source causing mass transfer. Dislocation emission by grain-boundary ledges, considered to be the principal grain-boundary sources, was modelled as being a thermally-activated mechanism. Penetration of an emitted dislocation was treated as being a soft collision. The overall macroscopic behavior was deduced by using a secant self-consistent scheme, with the material being viewed as the familiar two-phase composite in which the grain cores were inclusions with a behavior governed by dislocation-glide, and the matrix phase represented both the grain boundaries and the triple junctions and was controlled by the above-proposed dislocation-emission and penetration mechanism. Long-range stress fields, arising from the presence of grain boundaries, were accounted for by the critical glide resistance stress in the inclusions at 0K. Good agreement was found between measurement and prediction. A model[270], specifically aimed at explaining the breakdown of the Hall-Petch law, was based upon a continuum description involving an explicit finite-element formulation of the three-dimensional deformation of polycrystal grains, with the grain boundaries being considered to be surfaces of discontinuity - of finite thickness - embedded in that continuum. A phenomenological scheme then described the grain-boundary deformation mechanisms of sliding and opening accommodation. The model was tested by using atomistic results which were appropriate for copper at very high loading-rates. Tensile test simulations reproduced the inverse grain-size dependence of the macroscopic yield stress which was predicted by atomistic simulations, and especially the fact that the grain-size dependence of the yield stress was linearly related to the inverse square root of the grain size, as in the normal Hall-Petch law; but with a negative coefficient. The predictions were in good agreement with atomistic simulation results and with quasi-static experimental data. A further model[271] for predicting the inverse Hall-Petch effect in nanocrystalline material again treated the latter as consisting of two phases: grains of spherical or spheroidal shape and a grain-boundary phase. Deformation of the grains was assumed to be elasto-viscoplastic and to include a dislocation-glide mechanism, plus Coble and Nabarro-Herring creep. Deformation of the grain-boundary phase was taken to be the grain-boundary diffusion mechanism. An atomically-equivalent continuum model[272] was used to study viscoplastic behavior at the lower range of grain sizes that was normally modelled by using molecular dynamics methods. A two-phase composite model was described, into which 3 distinct inelastic deformation mechanisms were incorporated in order to build a general micromechanics-based scheme. The mechanisms included dislocation-related plastic flow in the grain interior, uncorrelated atomic motions within the grain-boundary zone and grain-boundary sliding at the interface between the grains and the grain-boundary zone. Viscoplastic behavior of the grain interior was modelled using a grain-size dependent constitutive equation, while the grain-boundary zone was modelled using a size-independent law. Grain-boundary sliding at the interface was treated as Newtonian flow. A rate-dependent work-hardening homogenization scheme was based upon an approach which went from elasticity to visco-elasticity according to the correspondence principle and then from visco-elasticity to viscoplasticity by replacing the Maxwell viscosity of the constituent phases by the respective secant viscosity. When applied to nanocrystalline copper having grain-sizes ranging from 5.21 to 3.28nm, under strain-rates of 2.5 x 10^8 to 1.0 x 10^9/s, the predicted results exhibited appreciable grain-size softening as well as strain-rate hardening. When applied to grain-sizes ranging from 40 to 2nm under the same high strain-rates, the flow stress first exhibited a positive slope and then a negative one in a Hall-Petch plot. Meanwhile the strain-rate sensitivity first increased and then decreased, while the activation volume first decreased and then increased. This suggested that the maximum strain-rate sensitivity and the lowest activation volume did not occur at the smallest grain size. Instead, like the maximum yield strength or hardness, they occurred at a finite grain-size. A higher yield strength, higher strain-rate sensitivity and lower activation volume on the positive side of the Hall-Petch plot were generally associated with an improved yield strength of the grain interior. Opposite trends, on the negative side of the plot, were associated with characteristics of the grain-boundary zone. Another two models[273] were aimed at explaining the inverse Hall-Petch effect and the stress-strain response of nanocrystalline copper during plastic deformation. In one, gradient plasticity together with an interface-energy term was used because of the predominant role played by nanoscale interfaces. The other involved a simplified gradient plasticity model, with no interface-energy term, coupled with wavelet analysis.

When molecular dynamics simulations were used[274] to model the yield strength of thermally-annealed nanocrystalline copper samples, the observed yield strengths at strain-rates of 10^{10} and 10^9/s scaled with the fractional number of grain-boundary atoms. This suggested the use of a new scaling behavior for the onset of plasticity in nanocrystalline materials; controlled not only by the grain size but by a combination of grain-size and degree of grain-boundary relaxation, as measured by the grain-boundary volume. Large-scale molecular dynamics simulations were used[275] to study the macroscopic yield behavior, at high strain-rates, of nanocrystalline copper having an average grain-size of 6nm. Simulations performed for strain-rates ranging from 10^9 to 8 x 10^9/s suggested that there was an asymmetry in the flow-stress values for tension and compression; with the nanocrystalline metal being stronger in compression than in tension. This asymmetry was very small at 10^9/s, but increased with increasing strain-rate. The predicted yield stresses and flow stresses under combined biaxial loading produced points which could described by an ellipse. The three-dimensional yield surface featured a cylinder which was symmetrical around the hydrostatic axis. It was concluded that a von Mises type yield criterion could be used to understand the macroscopic deformation behavior of nanocrystalline copper having a grain size which was in the range where the inverse Hall-Petch effect might be expected to occur.

Similarly to yield stress and hardness values, a Hall-Petch type of dependence is observed for the deformation twinning of copper at grain sizes greater than 1μm. When the grain size is decreased into the nanocrystalline range, the tendency to deformation twinning instead tends to increase and to exhibit an inverse grain-size dependence. This trend is again reversed at even smaller grain-sizes, and returns to a behavior of increased deformation-twinning difficulty with decreasing grain-size. This double-inversion behavior relative to the normal Hall-Petch grain-size dependence was demonstrated[276] for nanocrystalline copper films when deformed in tension at room temperature using low strain-rates. The non-monotonic grain-size dependence of deformation twinning was explained by modelling competing grain-size effects upon the emission of the initial partial dislocations, and the subsequent plane-to-plane promotion of partial dislocation slip.

Large-scale molecular dynamics simulations were used[277] to investigate the effects of microstructure and loading conditions upon the dynamic failure of nanocrystalline copper, together with the nucleation, growth and coalescence of voids. The simulated material had an average grain-size of 6 to 12nm, putting it into the inverse-Hall-Petch range. The assumed testing condition was uniaxial tension at a constant strain-rate of 4 x 10^7 to 10^{10}/s. The simulation results suggested that the evolution of voids involved two stages: The first stage was void nucleation and rapid linear initial growth of all of the individual voids. The second stage was the slower steady growth and coalescence of void aggregates and clusters. Changes in the void fraction depended strongly upon the strain rate, and were less dependent upon the grain size. Higher strain-rates required a larger plastic strain in order to nucleate voids, while larger grain sizes required a lower plastic strain in order to nucleate voids in the inverse-Hall-Petch range. Such simulations have also been used[278] to investigate the morphology-dependent mechanical properties of nanocrystalline copper. Degradation of the properties under thermal loading was investigated for high strain-rate deformation. The results showed that thermal loading of nanocrystalline material altered the grain-size behavior of mechanical properties.

A constitutive equation was proposed[279] which took account of important deformation mechanisms, including grain interior plasticity, grain-boundary diffusion and grain-boundary sliding. The stresses which were deduced from the constitutive equation closely matched experimental data on nanocrystalline copper. The model also predicted the variation in the strain-rate sensitivity parameter. Deviations from the Hall-Petch and inverse-Hall-Petch effects were also illustrated.

Experiment and simulation show that the introduction of nanoscale twins into nanocrystalline copper is an effective means for improving strength. Molecular dynamics, and the embedded-atom method, were used[280] to simulate a cell of [011]-textured microstructure with four hexagonal grains. The results showed that the strength and toughness could be increased by introducing twin boundaries into nanocrystalline grains. The nanotwins acted as obstacles to dislocation motion, thus leading to strengthening, as well as acting as sources of dislocation nucleation and thus contributing to the toughness. The improvement in the properties was sensitive to the distance between twin boundaries and grain boundaries, and exhibited a maximum at an intermediate distance. It decreased when twin boundaries were far away from or very close to grain boundaries, thus implying that the volume between the twin boundary and the grain boundary played an important role in determining the plasticity of nanocrystalline copper. The deformation behavior of the grains also depended upon their orientation with respect to the loading direction. Most importantly, grain-size refinement could lead to an inverse-Hall-Petch effect.

Large-scale three-dimensional molecular dynamics simulations[281] of the propagation of cracks at temperatures of between 1 and 500K in nanocrystalline copper specimens having average grain-sizes ranging from 5.34 to 14.8nm, showed[282] that intragranular fracture could occur within the grains at low temperatures. Plastic deformation around the crack tip was accommodated by dislocation nucleation and emission and fully-extended dislocations and deformation twinning were visible around the crack tip during fracture. Due to a higher concentration of stress ahead of the crack at lower temperatures, twinning deformation was easier to nucleate at the crack tip. These results also showed that decreasing the grain-size to below a critical value led to inverse-Hall-Petch behaviour due to the enhancement of grain-boundary involvement.

Atomistic simulations[283] have shown that anomalous, inverse-Hall-Petch, behaviour at small grain-sizes is not limited to the hardness and yield stress but is also exhibited by the cohesive energy and elastic constants. This effect was attributed to the greater relative concentration of grain-boundary atoms which existed when the grain-size was smaller. A model which considered separately the contributions made by atoms in the grains and grain boundaries, to the cohesive energy and elastic constants, agreed well with the simulation results. Calculations of the structural properties and elastic constants, and estimations of the hardness, showed that very different materials could exhibit qualitative similarities, such as a linear scaling (figure 18) of the fraction of non-crystalline atoms with respect to the reciprocal of the grain-size and similar scaling-laws for the cohesive

Materials Research Forum LLC
https://doi.org/10.21741/9781644900352

energy and elastic constants. Some quantitative differences, such as broader peaks in the pair-correlation function, could nevertheless lead to differing magnitudes of scaling coefficients. Theoretical analysis suggested universal scaling relationships for the properties of nanocrystalline materials as a function of the average grain size.

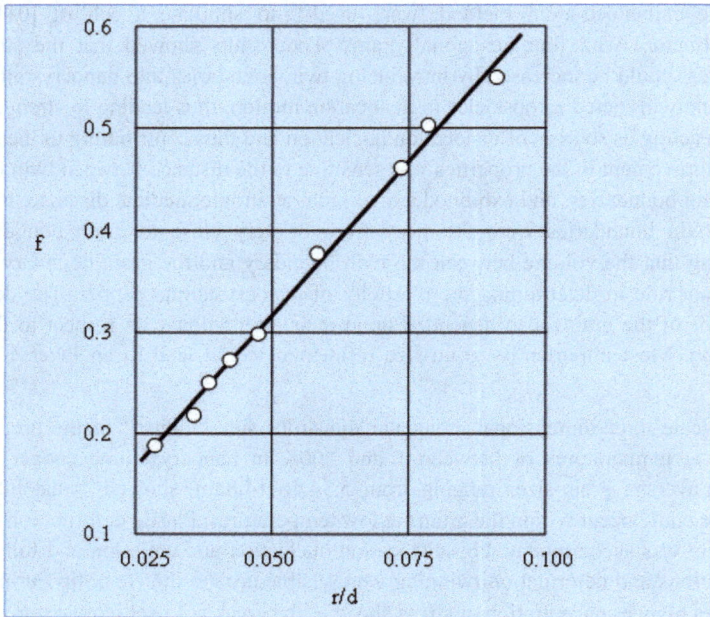

Figure 18. Fraction of copper atoms which are not part of the idealized crystalline structure, as a function of the ratio of atom size to grain size

During shaped-charge explosive compression the grain-size of the metal was reduced, from 30 to 80µm, to the sub-micron or nanometre level[284]. A model for nanocrystalline copper having grain-sizes of 7.17, 9.11, 12.55, 14.85, 18.38 or 22.48nm was based upon a Voronoi geometrical construction and was relaxed, within 100ps at 293K, to the equilibrium state. Tensile deformation was simulated by using molecular dynamics methods. The strain gradually increased to 0.2 at a strain-rate of 2×10^9/s. The results showed that the average flow stress was highest for a grain size of 14.85nm. When the grain size was 22.48nm, typical dislocation motion was observed and a huge number of dislocations was involved in the deformation. The number of dislocations decreased sharply for grain sizes of 14.85 and 9.11nm, and grain-boundary motion was visible. Dislocation motion dominated the deformation when the grain size was greater than 14.85nm. As the grain-size decreased to below 14.85nm, grain-boundary sliding and rotation became the predominant deformation modes. This change in mechanism was the main cause of softening, and thus of the inverse-Hall-Petch relationship.

The nano-indentation of monocrystalline and nanocrystalline copper was modelled[285] by using molecular dynamics simulations which were based upon the many-body tight-binding potential. The grain-size effect was analysed in terms of slip-vectors, atomic stresses, loading force and hardness. The inverse-Hall-Petch effect occurred for grain-sizes below 7nm. At grain-sizes smaller than 5nm, the equivalent stress quickly decreased and stress-induced grain-growth was observed during indentation. Grain rotation was the main cause of grain-coarsening at small grain-sizes. At larger grain sizes, dislocations were found on the {111} close-packed plane and on the {100} plane. The combined effect of grain size and texture upon the strength of nanocrystalline copper was studied[286] by using a plasticity-based model in which slip occurred via discrete slip events which involved individual dislocations that were emitted by the grain boundaries. A Hall-Petch relationship arose naturally for initially textured and non-textured materials, and the slope increased with increasing textural strength. This suggested that certain preferred orientations intensified the increases in strength which were associated with grain-size reduction. It was concluded that the existence of texture was too important to be neglected when exploiting the grain-size effect in order to increase the strength of nanomaterials. Molecular dynamics simulation of uniaxial tensile tests was also used[287] to study the effects of grain size and temperature upon the properties of quasi-two-dimensional nanocrystalline copper having grain sizes of 3 to 8nm. The results showed that grain growth was caused by grain rotation and led to strain-hardening at 300K for a grain-size of less than 5nm at 100K. At grain-sizes of up to 7nm, an inverse-Hall–Petch effect was observed in the yield stress. Molecular dynamics simulations were used[288] to study the mechanical properties of columnar nanocrystalline copper having a grain size of between 9.0 and 24nm. A melting and cooling method was used which generated samples that contained defects such as dislocations and vacancies within the grains. Uniaxial tensile deformation then revealed the existence of a critical mean grain size range, between 16 and 20nm, within which an inverse-Hall–Petch effect occurred. The main deformation mechanism involved a combination of dislocations and grain-boundary sliding. There was a sliding of planes which was generated by the motion of perfect edge dislocations that were absorbed by grain boundaries. The initial defects within the grains permitted this mechanism to occur.

The inverse-Hall-Petch effect has been observed[289] in pure copper sliding contacts at cryogenic temperatures. By kinetically limiting grain growth, it was possible to produce a quasi-stable ultra-nanocrystalline surface layer of reduced strength. *In situ* electrical contact resistance measurements were used to monitor grain-size changes at the interface. There was found to be a direct correlation between the surface grain-size and the coefficient of friction, thus confirming a theory which linked friction in pure metals to a transition from dislocation-controlled plasticity to grain-boundary sliding.

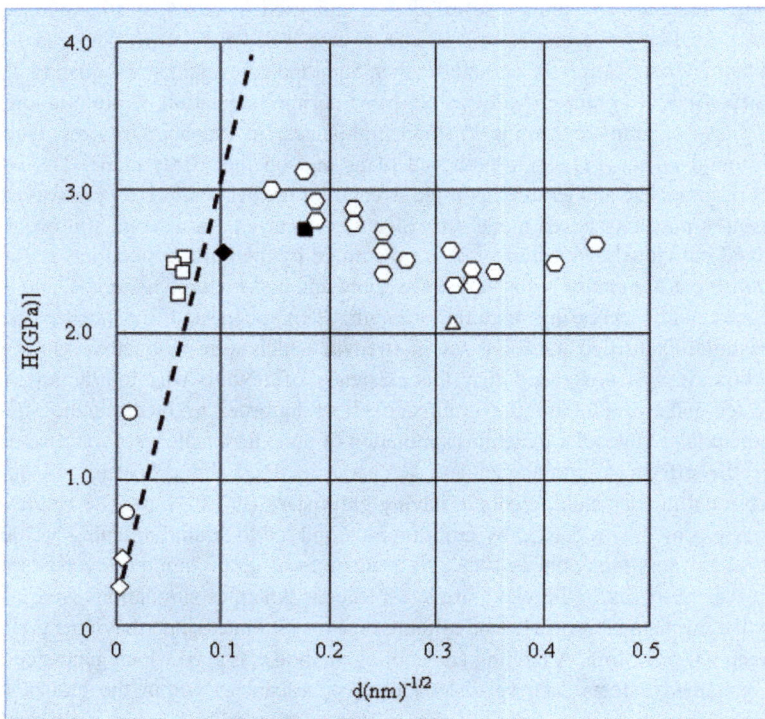

Figure 19. Hardness of Cu-30%Zn as a function of grain size. Dashed line: Meyers and Chawla[290], hexagons: present results, open diamonds[291], open circles[292], open squares[293], filled diamond[294], triangle[295], filled square[296].

When nanocrystalline Cu-Ni-P specimens, having average grain-sizes of 7, 10 and 24nm, were prepared[297] by electrodeposition, a breakdown of the Hall-Petch relationship was noted in both the tensile strength and the hardness data. Nanocrystalline Cu-30%Zn samples were produced[298] by high-energy ball-milling at 77K and room temperature, with cryomilled flakes being further processed via ultra-high strain high-pressure torsion or room-temperature milling in order to produce bulk artefact-free samples. Deformation-induced grain growth, and a reduction in the presence of twins, was observed in the high-pressure torsion consolidated samples. At grain sizes of less than about 35nm, the results deviated from the Hall-Petch behaviour (figure 19) and the strength became comparable to that of nanocrystalline pure copper. A high density of finely-spaced deformation nano-twins was formed, due to the low stacking-fault energy ($14mJ/m^2$) and the severe low-temperature plastic deformation. The present softening mechanism was likened to the twin-related softening behavior of nanotwinned copper.

Nickel

The unidirectional tensile deformation of nanocrystalline, 6 to 40nm, electrodeposited nickel revealed[299] a deviation from the Hall-Petch relationship and it was suggested that nanocrystalline materials could be described by a composite model. Further experiments showed that such nanocrystalline electrodeposits exhibited appreciable room-temperature creep. Grain-boundary sliding and diffusive matter-transport within the intercrystalline region were deduced to play an important role in the deformation of nanocrystalline materials. Electrodeposited dense ductile nanocrystalline samples[300] having various grain-size distributions had strengths which fell within the scatter-band of the overall Hall-Petch plot for nickel, although some large variations in yield strength, strain-hardening rate and elongation were linked to relatively small changes in the average grain size. A scatter in the elongation data was attributed to the formation of nodules and the presence of voids, while variations in the strength and strain-hardening were associated with changes in the grain-size distribution. The possibility of composite behavior was again invoked, together with the influence of confined dislocation motion. The effect of a reduced grain-size upon the wear resistance of pure nickel was explored[301] using Tabor abrasive-wear testing showing that, as the grain-size decreased from polycrystalline to nanocrystalline, the wear resistance appreciably increased in line with an increased hardness, while the volume-loss due to wear obeyed Archard's law. Following the wear tests, extensive plastic deformation of the polycrystalline material was observed. Plastic deformation markedly decreased with decreasing grain size, and became negligible at the smallest (13nm) grain-size. When the abrasion resistance was investigated[302] by using a nanoscratch technique and a ramped load, a breakdown in the Hall-Petch behaviour of the hardness was observed (figure 20) at the smallest (12 to 14nm) grain-sizes studied.

Changes in the abrasive wear behavior mirrored the changes in hardness, in spite of an apparent change in deformation mechanism at the finest grain-sizes.

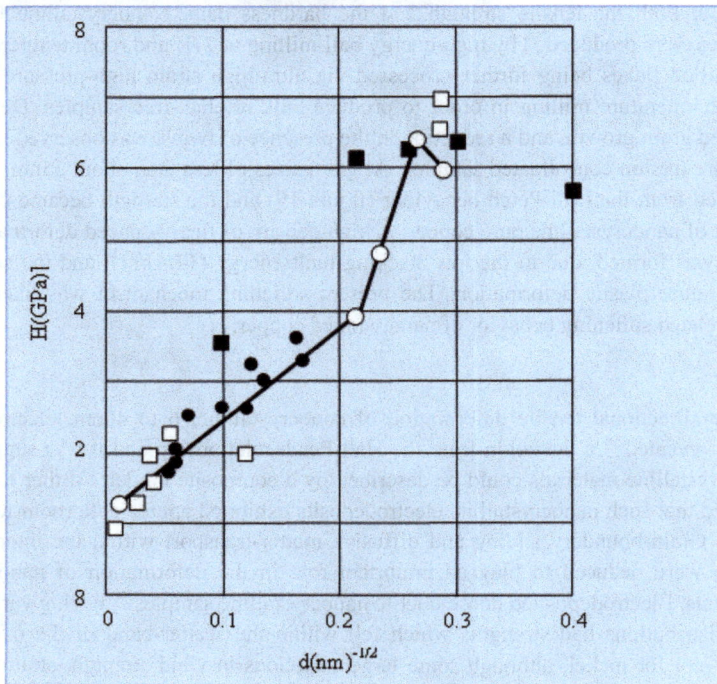

Figure 20. Hardness of nanocrystalline nickel as a function of grain size.
Open circles: present results, closed circles[303], open squares[304], closed squares[305,306]

Lattice-volume versus pressure data[307] revealed elastic softening in nanocrystalline material as compared with that of bulk nickel. An increased overall compressibility of nanocrystalline samples was attributed to a higher compressibility of a surface shell of nanocrystals; in agreement with molecular dynamics simulations. A distinction was drawn between a micro, or local, yielding due to high stress concentrations at grain-on-grain contact points and a macro, or bulk, yielding due to deviatoric stresses acting on the whole sample. Micro-indentation measurements revealed that annealing, at less than 40% of the reduced melting-point, hardened the nanocrystalline metal and led to an inverse-

Hall-Petch relationship (figure 21). The latter was explained in terms of impurity segregation to the grain boundaries.

Numerical simulations have been used[308] to investigate the competition between grain-boundary and dislocation-controlled deformation in nickel having grain sizes ranging from 4 to 32nm. A three-dimensional phase-field model was used to track the evolution of individual dislocations and boundaries, showing that the transition from Hall-Petch to inverse-Hall-Petch behaviour, with decreasing grain-size could not be attributed only to the grain-size but was also affected by grain-boundary energetics. The grain-size which corresponded to the maximum yield stress: that is, the one at which the transition from Hall-Petch to inverse-Hall-Petch behaviour occurred, decreased with increasing grain-boundary energy. When the cohesive energy of the grain-boundaries was sufficiently high, the Hall-Petch maximum was not observed within the grain-size range that was studied.

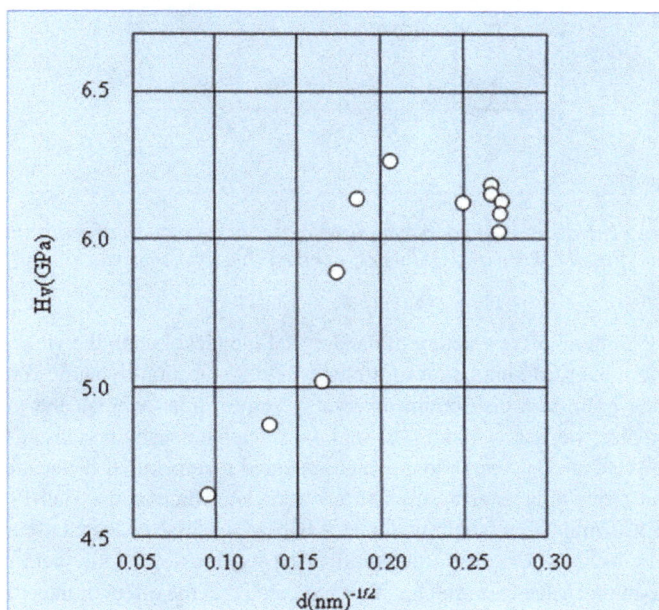

Figure 21. Hardness of nanocrystalline nickel at 623K as a function of grain size

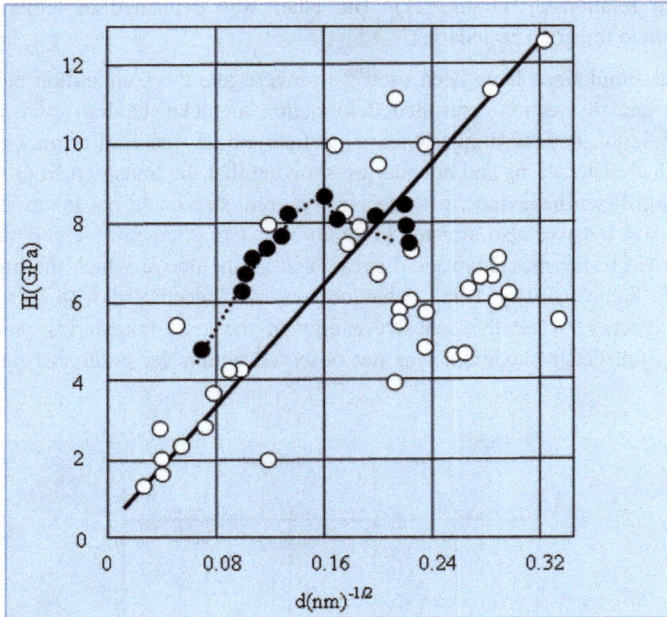

Figure 22. Hardness of nanocrystalline nickel as a function of grain size
Filled circles: present work, open circles: previous work

The effect of the initial microstructure of nanocrystalline nickel upon the yield stress was modelled[309] via numerical simulation of a phase-field dislocation-dynamics model. This showed that the grain-size distribution exerted a large influence upon the yield stress when the grain-size was below 32nm. The initial dislocation density was also a key factor governing the yield stress. Simulations which assumed a zero initial dislocation-density led to an almost size-independent stress-strain behavior, whereas the Hall-Petch effect was observed in simulations which assumed a non-zero initial dislocation-density. The results could be used to construct a probability-density function of the yield stress as a function of the microstructural variables. This indicated that the effect of uncertainty as to grain-size distribution and initial dislocation-density was more marked for smaller grain-sizes. The hardness of nanocrystalline nickel which would be measured by using a nanosized indenter was studied[310] by means of molecular dynamics simulations. It was found that the grain-size effect which was observed was different to that deduced from uniform deformation or from deeper indentation tests. The hardness could then exhibit only the inverse-Hall-Petch effect, with no Hall-Petch effect being observed for grain-sizes of up to 40nm. Grain-boundary absorption of localized strain was the main deformation mechanism which operated when the indenter-size and the depth were both nano-sized. The area of the plastic zone which was generated beneath the tip depended strongly upon the grain-boundary density. A sample possessing a small grain-size then had a larger plastic area, leading in turn to a lower hardness. Large-scale molecular dynamics simulations were used[311] to investigate the low-temperature friction behavior of material having grain-sizes ranging from 4 to 11nm. A reduction in the coefficient-of-friction occurred with decreasing grain-size down to 5nm, at which point the trend was reversed. Molecular dynamics simulations[312] of the compressive behavior of nanotwinned polycrystalline nickel nanowires predicted the occurrence of an inverse-Hall-Petch behaviour as a function of twin-boundary spacing. This was explained in terms of a change in the predominant intra-granular deformation mechanism.

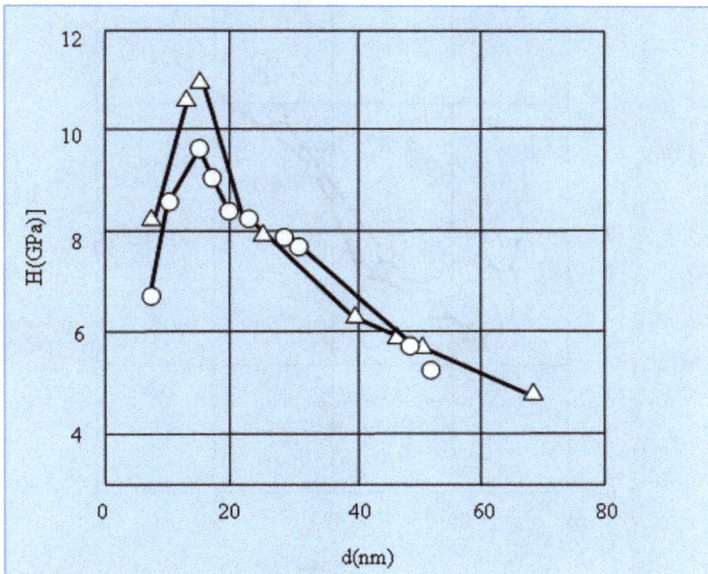

*Figure 23. Hardness of nanocrystalline nickel as a function of grain size
Circles: (200) preferred orientation, triangles: (111) preferred orientation*

Materials Research Forum LLC
https://doi.org/10.21741/9781644900352

Indentation-creep and stress relaxation testing of rolled and annealed nanocrystalline nickel were used[313] to study the effect of microstructural changes upon plastic deformation. The dislocation density increased, with increasing rolling strain, to reach a maximum at 20% strain. This was followed by a decrease at greater strains. The dislocation density decreased markedly with increasing annealing temperature. An inverse-Hall-Petch effect was observed in specimens having a grain-size of less than 40nm (figure 22). In rolled nanocrystalline nickel, the creep strain-rate and rate-sensitivity first increased and then decreased. In annealed nickel, they decreased monotonically. With increasing grain-size, the creep activation-volume first anomalously decreased, and then began to increase. This behaviour was different to that of coarse-grained specimens. The results were explained in terms of dislocation annihilation and emission at grain boundaries.

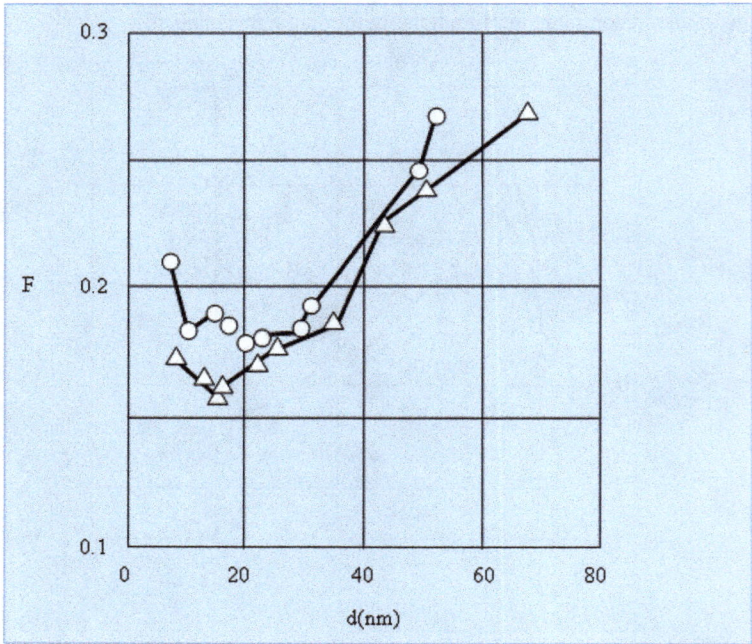

Figure 24. Coefficient-of-friction of nickel as a function of grain size
Circles: (200) preferred orientation, triangles: (111) preferred orientation

In nickel films which were produced by using the super-critical electroplating process the preferred orientation and grain size were determined by the nature of the substrate (brass, copper), by the reaction time and by the chamber-pressure, with (111), (111)+(200) and (200) preferred orientations being observed[314]. As expected, the mechanical and tribological properties improved with decreasing grain size. But here again, at grain-sizes of less than 30nm, such improvements were stopped or even reversed by the appearance of the inverse-Hall-Petch effect (figures 23 and 24); here attributed to internal stress. It was noted that (111) type films tended to exhibit superior mechanical and tribological properties because – it was claimed – that on (111) crystal planes the nickel atoms accumulated in the most compact form and a stronger binding energy existed between the atoms. It was concluded that, if the grain-size is in the nano-scale range, the mechanical and tribological properties are under the control of the film's anisotropy.

Nanocrystalline NiAl powder with grain-sizes ranging from 7 to 80nm was produced by mechanical alloying and then variously heat-treated. Depending upon the treatment, the material could exhibit an inverse-Hall-Petch behavior[315] at grain-sizes below 20nm (figure 25) while, in other cases, the hardness increased until it levelled off at 7nm.

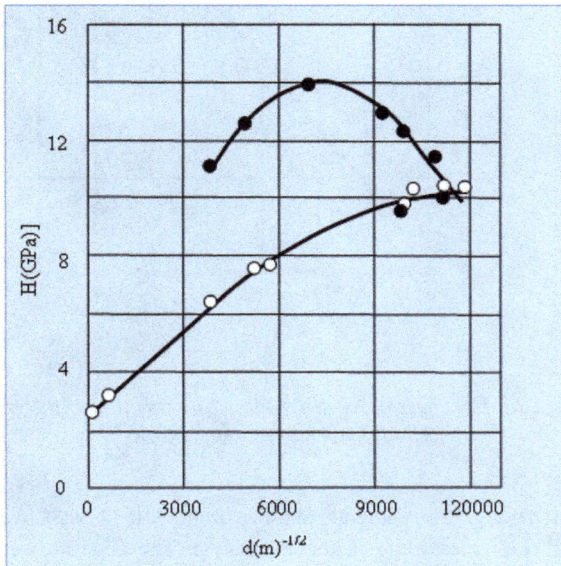

Figure 25. Microhardness of NaAl as a function of grain size
Open circles: as-milled, filled circles: heat-treated

When Ni-Al-Cr alloy having an initial grain-size of approximately 60μm was deformed in torsion to a strain of about 7 at room temperature, the grain size was reduced to about 34nm. The post-straining alloy contained dislocations which were associated with lattice distortions, with no ordered Ni_3Al. The ultra-fine grains were very stable at high temperatures, and a grain-size of less than 100nm could be retained during annealing at up to 900K. The stability of the grains was attributed to the formation of an ordered Ni_3Al-based phase during annealing at 650 to 700K. The presence of this phase led to a negative slope of the Hall-Petch law (figure 26)[316].

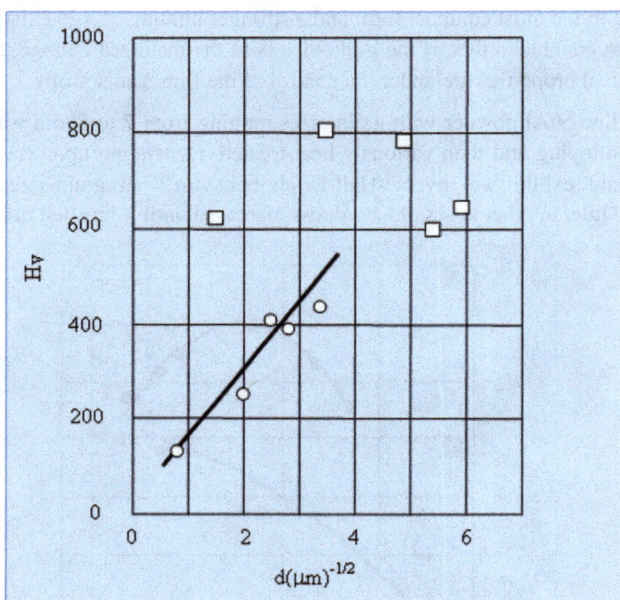

Figure 26. Hardness of Ni-Al-Cr alloy as a function of the grain-size
Circles: nickel, squares: Ni-Al-Cr

Electrodeposits having iron contents ranging from 7 to 31wt% were obtained[317] by varying the Ni^{2+}/Fe^{2+} mass ratio in the electrolyte. The deposits were nanocrystalline, with an average grain-size of between 20 and 30nm. The grain-size decreased with increasing iron content; especially for short-term electroplating. Increasing the electroplating time had no appreciable effect upon the grain size. The microhardness exhibited normal Hall-Petch behaviour, with a maximum of 762H_v for a Ni^{2+}/Fe^{2+} mass ratio of 9.8. Thick deposits, having grain-sizes of less than 30nm, of nanocrystalline nickel and of alloys containing up to 28wt%Fe were again produced[318] by electrodeposition at rates greater than 100μm/h. The grain-sizes, macrotextures and microtextures depended strongly upon the amount of iron which was co-deposited with the nickel. With decreasing grain-size, the hardness initially increased in accord with the normal Hall-Petch relationship but, beginning at grain-sizes of about 18nm, there was softening at the smallest grain-sizes (figure 27).

Nano-indentation studies of electrodeposited nanocrystalline nickel and nickel-iron alloys revealed[319] a transition from normal to inverse-Hall-Petch behaviour (figure 28). There was no significant effect of the grain size upon the Young's modulus of nanocrystalline nickel and nickel-iron alloys having grain-sizes greater than 20nm. The moduli of nanocrystalline nickel and nickel-iron alloys, for grain-sizes less than 20nm, were slightly low when compared with those of randomly oriented polycrystalline equivalents. The trend as a function of decreasing grain-size was consistent with the composite-models which considered the effect of intercrystalline defects. The texture was considered to exert an influence upon the measured Young's modulus, and this was used to explain some of the behaviour observed over the entire range (9.8 to 81nm) of grain-sizes.

Nickel-iron coatings containing up to 75wt%Fe were produced[320] by electrodeposition at room temperature from an additive-free electrolyte, using current-densities ranging from 1 to 5A/dm^2. Hydrogen evolution affected the alloy structure, especially the composition limits of the γ-phase. Solid-solution effects contributed greatly to the strength of γ-phase alloys over the composition range of 5 to 25%. With decreasing grain-size, increasing levels of internal stress and a decreasing stiffness led to a marked softening of nanocrystalline γ-phase alloys having iron contents greater than about 25%. The effect of iron additions (0, 18.5, 28.5 or 43wt%Fe) to nickel upon nanostructure retention during annealing (450C, 16h) was studied[321]. The integral-breadth method revealed a decrease in grain-size with increasing iron content. The strain-rate sensitivity exponent was higher (0.10803) for 18.5wt%Fe alloy during nano-indentation, while 0, 28.5 and 43wt%Fe alloys exhibited relatively low strain-rate sensitivity exponents of between 0.02069 and 0.10803. The Hall-Petch relationship was followed by alloys with up to 18.5wt%Fe, while an inverse-Hall-Petch relationship was followed by alloys having higher iron contents. The microstructures of nanocrystalline nickel-iron alloys, deposited from electrolyte solutions, were controlled[322] by varying the current-density so as produce specific grain-sizes, iron contents, lattice strains and textures. The grain-sizes ranged from 6 to 17nm, and the textures were of {111} and {200} orientation. The 0.01%-offset bending yield strengths, for the above grain-size range clearly obeyed the Hall-Petch law.

Materials Research Forum LLC
https://doi.org/10.21741/9781644900352

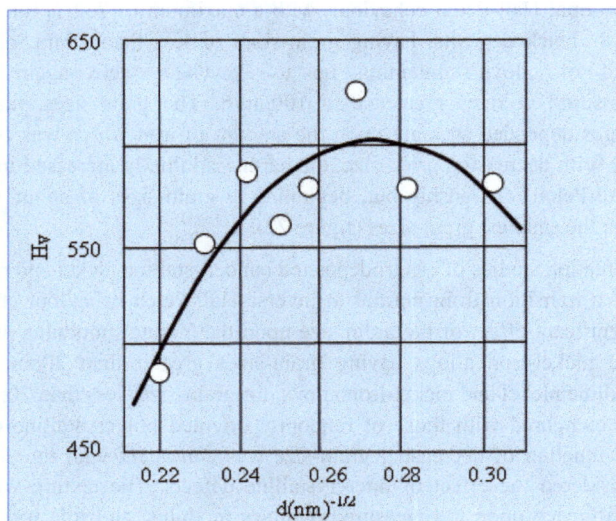

Figure 27. Hardness of electrodeposited nickel-iron alloy as a function of grain-size

There was a lower work-hardening and lower strain at fracture with decreasing grain-size. The Young's modulus, indentation modulus, indentation hardness and bending strength were essentially unaffected by the microstructure. A study[323] of electrodeposited bulk nanocrystalline Ni–Fe and Ni–W alloys was aimed at predicting grain-boundary relaxation; a possible rival factor to that of dislocation pile-up in explaining grain-size related hardening. Low-temperature heat-treatment promoted grain-boundary relaxation and thereby increased the hardness by between 0.07 and 0.74GPa. The increase in hardness decreased with increasing orientation index for the (200) plane, and it was concluded that electrodeposited nickel alloys having an orientation index greater than 3.0, for the (200) plane, would not exhibit grain-boundary relaxation strengthening. This was because those alloys did not possess a non-equilibrium grain-boundary structure, even in the as-deposited state.

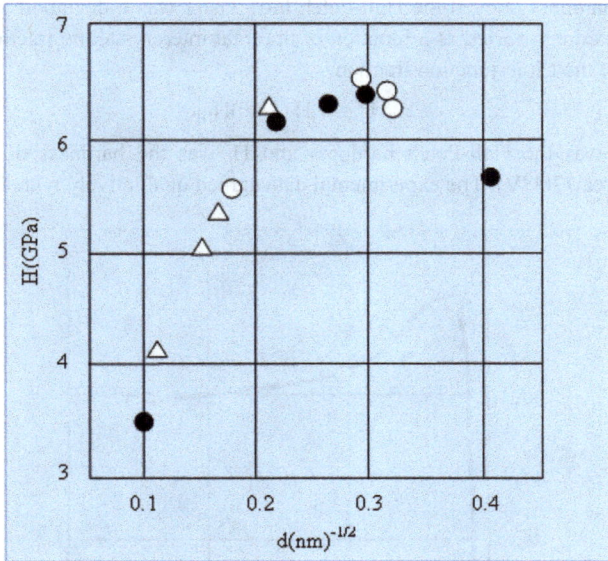

Figure 28. Hardness of nickel-iron as a function of grain-size. Open circles: Ni-Fe present work, triangles: nickel current work, filled circles: nickel previous work

In electrodeposited nanocrystalline Ni-Fe-W alloys[324], the crystallite size decreased with increasing current-density due to an increase in the amount of alloying element. A Ni-23Fe-1.3at%W alloy, deposited using $0.1A/cm^2$, exhibited the greatest hardness ($563H_V$) and best wear-resistance. The results obeyed the normal Hall-Petch relationship for crystallite sizes above 12nm and an inverse-Hall-Petch relationship below 12nm (figure 29). The wear-resistance increased with decreasing crystallite size in the direct Hall-Petch range and decreased in the inverse-Hall-Petch range (figure 30).

Nanocrystalline layers of Ni-Mo alloy were electrodeposited[325], using direct-current, from citrate-ammonia solutions. The grain-size decreased with increasing molybdenum content, and was smaller for layers deposited at a pH of 9.5 than for those deposited at a pH of 8.5. The maximum microhardness was close to $800H_v$ for 17wt%Mo layers. At higher molybdenum contents there was a softening due a deviation from Hall-Petch behaviour at small grain-sizes (figure 31). A rough estimate of the hardness could be made by using an empirical grain-size dependence. For mean grain-sizes greater than

10nm, the hardness obeyed the Hall-Petch law. There was a deviation at smaller grain-sizes. The hardness varied as a function of the total intercrystalline fraction, V_i, rather as a function of the triple-junction fraction,

$$H = V_iH_i(1-V_i)H_{HP}$$

where H_{HP} was the Hall-Petch hardness and H_i was the hardness of the amorphous material (circa 770HV). The experimental data agreed qualitatively with this fit.

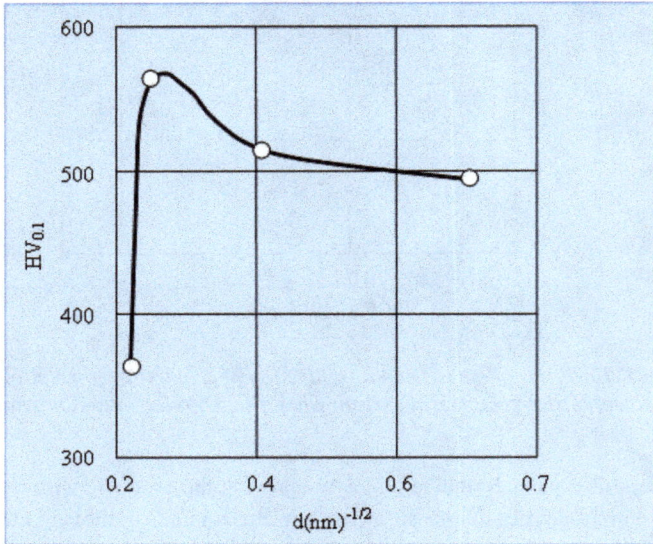

Figure 29. Hardness of Ni–Fe–W alloy as a function of crystallite size

In other work[326], nanocrystalline Ni-Mo coatings were deposited from citrate-ammonia solution by using rotating (0 to 640rpm) low-carbon steel discs as the cathode, electrolyte pH-values ranging from 4 to 10 and temperatures of between 20 and 60C. The optimum conditions were an electrolyte pH-value of above 7, a temperature of between 20 and 40C and a rotation-rate of 260 to 640rpm. The molybdenum acted as a grain-refinement modifier and also had a marked effect upon the mechanical properties of the layers. At molybdenum contents of less than 16wt%, and at an associated crystallite size of 7nm, a loss of strength due to the inverse-Hall-Petch relationship was again observed, together with a decrease in the wear-resistance.

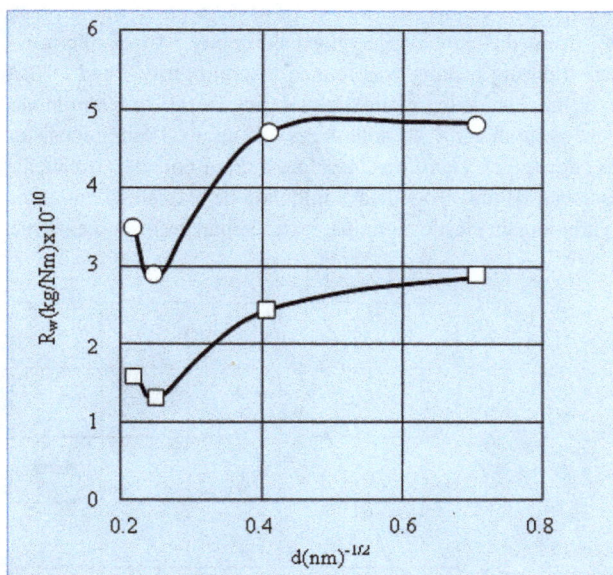

*Figure 30. Wear rate of Ni–Fe–W alloy as a function of crystallite size
Squares: 5N load, circles: 10N load*

Annealing at 600C of Ni-Mo-B alloys containing between 27 and 31.5at%Mo plus 10at%B produced[327] a granular phase consisting of face-centred cubic nanocrystallites with an average grain-size of 15 to 25nm, depending upon the chemical composition. The nanocrystalline phase consisted of grains of a face-centred cubic solid solution of molybdenum and boron in nickel, dispersed in an amorphous matrix. The lattice parameters of the nanocrystallites depended upon the alloy composition and the isothermal-annealing temperature. During this annealing, molybdenum and boron atoms diffused from the solid-solution lattice and into the amorphous surroundings. The stability of the nanocrystalline structure was governed by the thermal stability of the amorphous matrix, the crystallization temperature of which increased with isothermal annealing time due to enrichment by the boron and molybdenum. The alloy became harder as the grains grew in size. The relationship between the hardness and the grain size was contrary to the Hall-Petch relationship, and this was attributed to hardening of the amorphous matrix due to changes in its composition.

Pulse-electrodeposited nanocrystalline Ni-1.5wt%P with an initial grain-size of 3nm was studied[328] by using differential scanning calorimetry. An exothermic calorimetry peak which appeared during heating was related to grain growth and Ni_3P formation. Nano-indentation of samples having grain-sizes ranging from 3 to 50nm indicated a breakdown in Hall-Petch strengthening at grain-sizes of up to 10nm; consistent with previous observations (figure 32). There was therefore a grain-boundary weakening range at grain-sizes of less than 10nm. The data could not be explained in terms of micro-strain relaxation, variations in elastic modulus, texture-changes or duplex-structure formation.

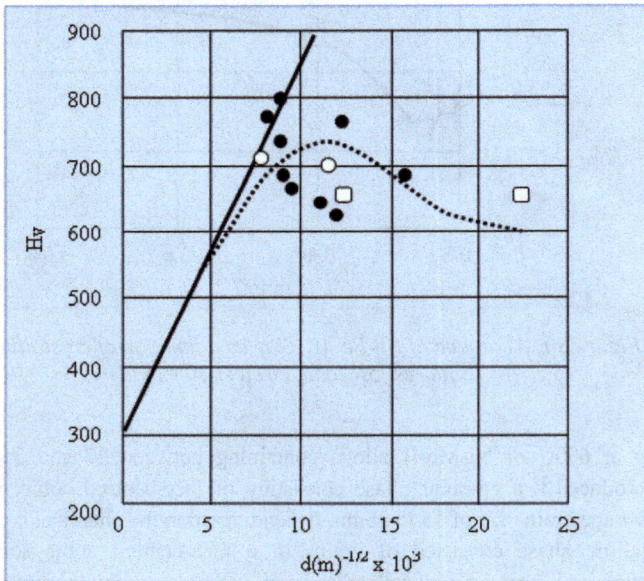

Figure 31. Microhardness of Ni–Mo deposits as a function of mean grain-size. Solid line: Hall-Petch relationship, dotted line: open circles: 11 to 13wt%Mo, filled circles: 16 to 20wt%Mo, squares: 25 to 30wt%Mo

Nanocrystalline $Ni_{43}Ti_{38}Al_{19}$ thin films with a thickness of 600nm, and a grain-size of 12 to 28nm contained $L2_1$-Ni_2TiAl, Ni_3Ti precipitates and some B2-NiTi. Films which contained the $L2_1$ phase exhibited a notable pseudo-elasticity which depended markedly upon the grain-size. The pseudo-elasticity decreased gradually with increasing grain-size and indentation-depth, and the highest pseudo-elasticity recovery-ratio was 92.7%. Upon

Materials Research Forum LLC
https://doi.org/10.21741/9781644900352

decreasing the grain size from 28 to 12nm, the hardness first increased and then decreased (figure 33), in accord with an inverse-Hall-Petch effect[329].

Electrodeposited nickel alloys containing about 13at%W, and having grain-sizes of less than 10nm, exhibited[330] a higher hardness and scratch resistance than did fine pure nanocrystalline nickel alloys, even though the effect of solid-solution strengthening due to the tungsten addition had been expected to be almost negligible. The improved properties were therefore attributed to the finer scale of the multicomponent nanocrystalline alloys, and particularly to the possibility that alloying had suppressed any inverse-Hall-Petch effect at minute grain-sizes. The overall results smoothly connected the hardness versus grain-size data covering the range from nanocrystalline to amorphous nickel-based alloys. Tensile testing of electrodeposited nanocrystalline Ni-W specimens, having grain-sizes of 20, 12, 8 or 5nm, showed[331] that a transition from a normal to inverse-Hall-Petch behaviour was a typical feature of such alloys (figure 34).

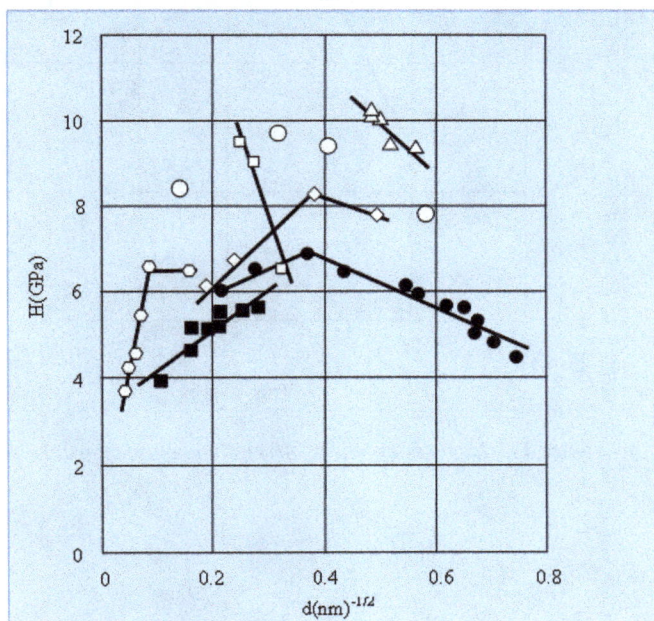

Figure 32. Hardness of electrodeposited Ni-P nanocrystalline alloys as a function of grain-size. Open circles: present work (1.5P), triangles: previous work (8.7 to 13.9P), open squares: previous work (1.2P), diamonds: previous work (2P), filled circles (P), hexagons: previous work (4.4P), filled squares: previous work (P)

In similar work[332,333], the crystallite-size decreased with increasing current-density due to an increase in the tungsten content. Alloys which contained 9.33at%W, and which were plated at 75C, exhibited a maximum hardness of 638H$_V$. Alloys which were plated at 75C obeyed the normal Hall-Petch law while alloys which were plated at 85C obeyed an inverse-Hall-Petch relationship for crystallite sizes below 15nm (figure 35). The wear-resistance of alloys which were plated at 75C increased due to an increase in the hardness with decreasing crystallite size to 20nm. Brittle fracture of the coating occurred below 20nm. The wear-resistance of alloys which were plated at 85C increased with decreasing crystallite size in the normal Hall-Petch region, and decreased in the inverse-Hall-Petch region.

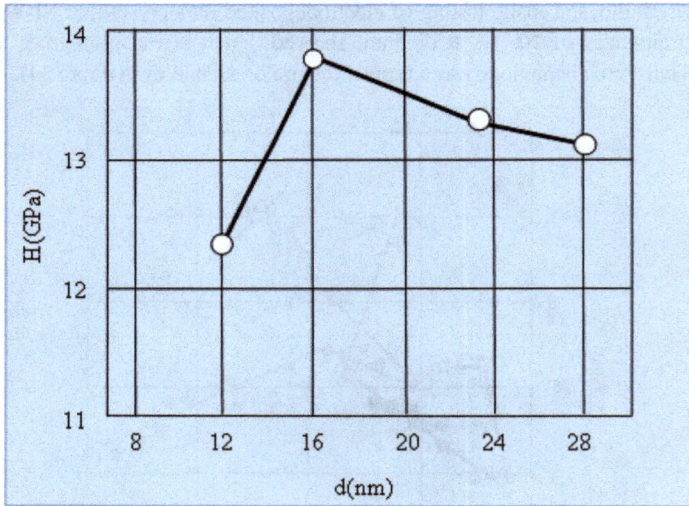

Figure 33. Hardness of Ni$_{43}$Ti$_{38}$Al$_{19}$ as a function of grain-size

Materials Research Forum LLC
https://doi.org/10.21741/9781644900352

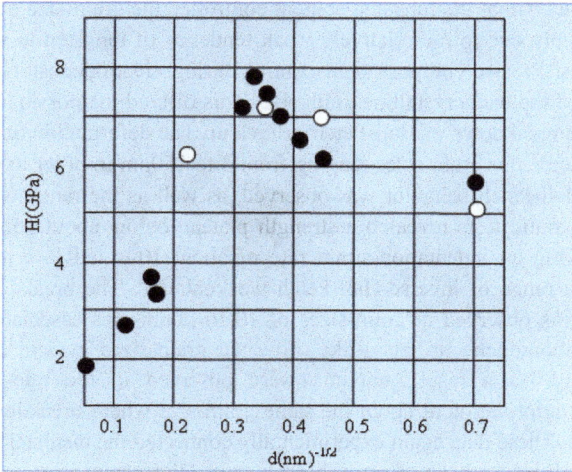

Figure 34. Hardness of electrodeposited Ni–W alloys as a function of grain-size
Open circles: present work, filled circles: previous work

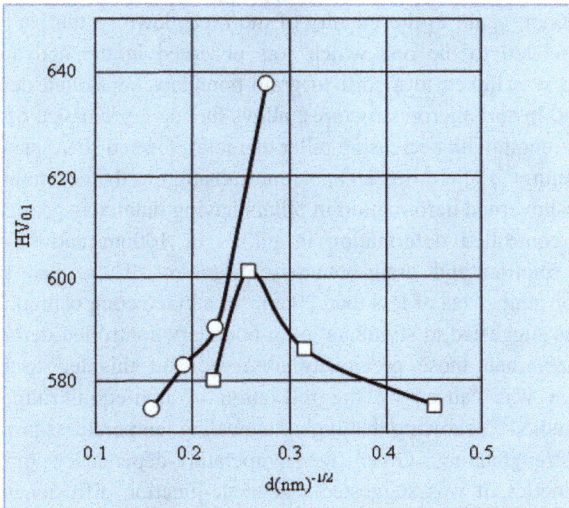

Figure 35. Hardness of electrodeposited Ni–W alloy as a function of crystallite size
Circles: deposited at 75C, squares: deposited at 85C

It was again noted[334] that the tungsten content controlled the grain-size over a range of 2 to 140nm, probably due to the relatively weak tendency of tungsten to segregate to the grain boundaries. Precise compositional control during electrodeposition could lead to precise control of the nanocrystalline grain size. This offered the possibility of clarifying the causes of the breakdown of Hall-Petch behaviour. The deformation of nanocrystalline alloys was studied[335] for grain-sizes ranging from 3 to 150nm in order to cover the range where normal Hall-Petch behavior was observed, as well as the range where deviations occurred. Quasi-static tests revealed a strength plateau below about 15nm, while high-rate tests involving an indentation-strain rate of about 10^3/s led to a marked strength maximum and a range of inverse-Hall-Petch weakening[336]. The breakdown in strength scaling which was observed at grain-sizes of 10 to 20nm was associated with a clear transition to inhomogeneous glass-like flow at grain-sizes which approached the amorphous limit. As a result, maxima were observed in the rate- and pressure-dependences of deformation at about the same grain-size where breakdown of the Hall-Petch law began. These data again experimentally connected the mechanical properties of nanocrystalline alloys to those of amorphous metals. High-temperature nano-indentation was used[337] to estimate the activation enthalpy for the deformation of nanocrystalline alloys having grain-sizes of between 3 and 80nm. Thermal softening was less marked for smaller grain-sizes and the activation enthalpy appeared to have an inflection at grain-sizes of 10 to 20nm; again in the vicinity of the breakdown of the Hall-Petch law. The inflection was related to the one which was observed in the activation volume for deformation and was linked to a shift to grain-boundary controlled deformation at the finest grain sizes. In sub-micron structures, alloys having a grain-size of 60nm exhibited a lower tensile strength with decreasing pillar diameter, formed shear bands and exhibited mechanical twinning[338]. There was an apparent transition in the deformation mechanism, from dislocation-governed deformation in pillars having diameters greater than 100nm to grain-boundary controlled deformation in pillars of 100nm and below. There also occurred grain rotation and grain-boundary migration. These had previously been observed only for grain-sizes of less than 20nm, for a given composition. The presence of free surfaces was suggested to stimulate grain-boundary controlled deformation at much greater grain-sizes than those previously observed and this led to lower strengths. Hardening which was caused by the relaxation of non-equilibrium grain-boundary structures was studied[339], showing that higher annealing temperatures produced faster and more marked strengthening. Given the temperature-dependence of the relaxation-strengthening kinetics, it was suggested that triple-junction diffusion was the limiting factor in the rate of removal of excess grain-boundary defects. The degree of relaxation-strengthening was studied for a range of grain-sizes which encompassed the breakdown of normal Hall-Petch behaviour. An apparent maximum in the hardening effect was found at a grain size below 10nm. The apparent activation volume for plastic deformation remained unaffected by annealing, for grain-sizes down to some 10nm, but increased with annealing in the case of the very finest grain-sizes. This again suggested that there was a change in the predominant deformation mechanism. When alloys were prepared by electrodeposition from an additive-free citrate ammonium bath, varying the conditions led to tungsten contents of up to 18at% and grain-sizes ranging from 5 to 650nm[340]. The incorporation of hydrogen, oxygen, carbon or nitrogen depended upon the deposition conditions, and the texture changed in the order: {110}, no texture, {111}. The Hall-Petch relationship was affected by the presence of light elements, by the crystallographic texture and by the type of grain boundaries. The relationship between grain-size and flow stress was directly related to the solute content and to changes in internal stress with grain-size. A competition between grain-boundary shearing and dislocation-emission at grain boundaries was suggested to explain the data. In the very latest studies[341], nanocrystalline and amorphous materials having tungsten contents of up to 34at% were produced by electrodeposition, with grain-sizes ranging from 25nm to fully amorphous. Solid-solution strengthening was observed; something which had been absent from alloys containing less than 13at%W. A Hall-Petch plot indicated a critical grain-size of 5nm (figure 36). This critical grain-size, and those of other Ni-W alloys, could be closely fitted by an equation of the form,

$$d_a/d_p = [(D_W/D_{Ni})(1-c) + c]^{2/7}$$

where d_a was the critical grain-size of a Ni-W alloy, d_p was the critical grain-size of pure nickel, D_W was the grain-boundary diffusivity of tungsten in nickel, D_{Ni} was the grain-boundary self-diffusivity of nickel and c was the atomic fraction of tungsten.

By using a Ni-W plating bath which contained dimethylamine borane as a boron source and citrate and glycine as complexing agents, Ni-W-B films having tungsten contents of up to 19at% and boron contents of up to 18at% were prepared by adjusting the tungstate concentration, glycine concentration and current density. The hardness of the nanocrystalline Ni-W-B films was proportional to the square root of the grain size, as expected, and attained a maximum of about $850H_v$ but, as usual, in the amorphous region or when the grain-size was less than 2 to 3nm the hardness again decreased[342].

Materials Research Forum LLC
https://doi.org/10.21741/9781644900352

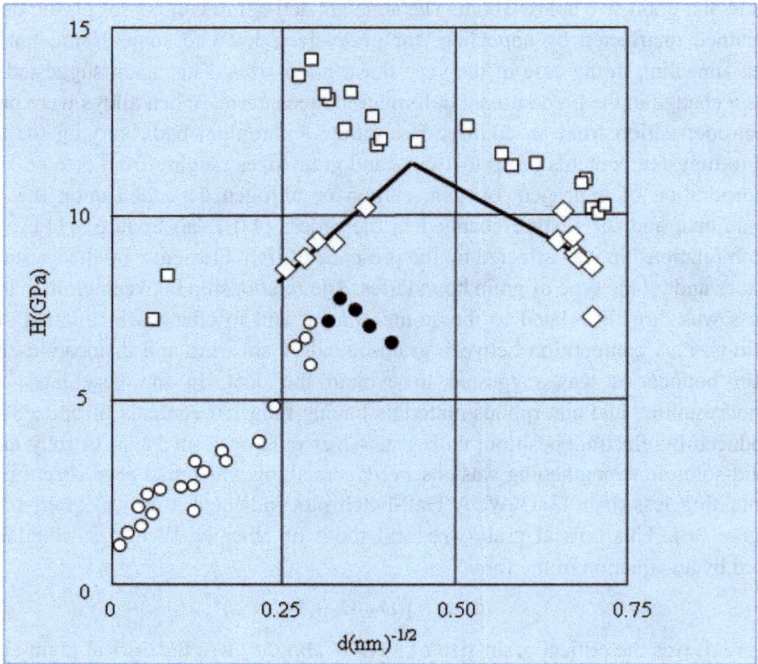

Figure 36. Hardness of electrodeposited Ni-W alloys as a function of grain size. Diamonds: present work (15.5 to 23.0at%W), filled circles: previous work (11.6 to 13.2at%W), open circles: previous work (pure electrodeposited nickel), squares: previous work (annealed 25at%W)

Samples of nanocrystalline $NiZr_2$, prepared by crystallizing an amorphous form, exhibited[343] a lamellar nano-twinned structure having a $(1\bar{1}0)$ direction at scales ranging from a few interatomic distances to a few nanometres. The microhardness of the single-phase nanotwinned material was greater than that of the originally amorphous material. For average grain-sizes ranging from 19.1 to 93.9nm, the hardness obeyed the normal Hall-Petch law but, at grain-sizes of less 19.1nm, it deviated from that relationship.

Zinc

Turning now to hexagonal close-packed materials, specimens of (104)-oriented monocrystalline zinc were bombarded[344] with 500keV carbon ions under ultra-high vacuum to doses of between 3.94 x 10^{14} and 1.30 x 10^{16}/cm², leading to the formation of nano- and sub-micron rods, clusters and other structures. The surface roughness increased with ion dose. The carbon content ranged from 22.86 to 31.20wt%, while the oxygen content was essentially constant at 10.16wt%. The crystallite size and lattice strain decreased with increasing ion dose, and a closely linear relationship existed between the crystallite size and lattice strain. The surface hardness decreased with ion dose, and obeyed an inverse-Hall-Petch law.

Samples of Zn-4wt%Al casting alloy, prepared[345] using sand-casting, die-casting and high-energy cryogenic ball-milling, were compared. In the case of cast samples there was an increase in strength in going from coarse-grained sand-cast microstructures to fine-grained die-castings. There was a decrease in strength and an increase in ductility when a cast structure was broken up by high-energy cryogenic-ball milling so as to give a uniform ultra-fine grain-size. Ultra fine-grained structures which were produced by cryogenic ball-milling and then subjected to isothermal heat-treatments obeyed the normal Hall-Petch law.

Pulsed laser deposition of zinc plus the introduction of a few monolayers of tungsten was used[346,347], to control the grain-size of nanocrystalline composites, that of the zinc being governed by the amount of zinc and the substrate temperature. Zinc islands, possessing a lower surface energy, nucleated on tungsten layers having a high surface energy and zero solubility in zinc. Nanocomposites having grain-sizes ranging from 30 to 6nm were thus produced. Their hardness roughly obeyed the Hall-Petch relationship. There was however the usual decrease below a critical grain-size of about 11nm which, in this case, was explained in terms of grain-boundary deformation and/or sliding. The role played by tungsten in grain-boundary deformation was of particular significance.

Magnesium

Magnesium has been processed[348] by using the hot-extrusion of ball-milled powder with a grain-size of 120μm to 60nm. In the microcrystalline product, the usual dislocation interactions with grain and/or twin boundaries led to a normal Hall-Petch relationship between flow stress and grain-size. The Hall-Petch slope became negative however for grain-sizes of less than 1μm because of the intervention of deformation due to twinning. When the grain size became less than 100nm, the twinning was also markedly reduced and the proportion of grain-boundary sliding providing plastic deformation increased; again leading to inverse Hall-Petch behaviour. There was also a negligible strain-

Materials Research Forum LLC
https://doi.org/10.21741/9781644900352

hardening rate, a relatively high strain-rate sensitivity and a low activation-volume. A novel method for avoiding the inverse-Hall-Petch effect is to combine nanocrystalline and amorphous materials[349]. Thus a magnesium alloy has been prepared so as to consist of nanocrystalline cores embedded in glassy amorphous shells. The crystalline phase consists of almost dislocation-free 6nm grains and blocks the propagation of most localized shear bands. Any bands which persist are impeded by rotation and division of the grains. The resultant dual-phase material possesses a close-to-theoretical strength of 3.3GPa.

The deformation behaviour of AZ31 alloy, deformed using equal channel angular extrusion, was investigated[350]. This revealed an inverse-Hall-Petch relationship between the yield stress and the grain-size. The tension-compression asymmetry was also weakened. The extruded material exhibits a strain-rate sensitivity, and the strain-rate sensitivity factor increased with increasing processing temperature. The average grain-size of as-extruded alloy was 20μm, with a typical ring basal texture, and the main room-temperature deformation modes were basal slip and twinning; thus resulting in a marked tension-compression asymmetry. The average grain size was about 2μm for the extruded alloy, with a relatively random texture. This led to a remarkable decrease in twin volume fraction under compression, and thus the tension-compression asymmetry was weakened. The calculated activation energy resembled that for grain-boundary diffusion in magnesium alloys, and thus implied that grain boundaries played an important role in deformation, and that there was a possible grain-boundary sliding contribution to the inverse Hall-Petch relationship and strain-rate sensitivity in these extruded alloys.

The mechanical behaviour of nanocrystalline magnesium and Mg-Al amorphous alloy has been modelled[351] by means of molecular dynamics simulation, revealing that the deformation mechanism of nanocrystalline magnesium is clearly affected by the amorphous boundary spacing and by temperature. That is, its strength increased with decreasing amorphous boundary spacing; leading to Hall-Petch behaviour at 10 and 300K. A stress plateau and subsequent elastic softening plus linear plastic strengthening were observed when the amorphous boundary spacing was less than 8.792nm at 10K. It was suggested that the amorphous boundary contributed to dislocation emission and absorption, but no second stress-peak was observed at 300K. Here the flow stress in the plastic stage was almost constant. The simulation showed the emergence of new grains, associated with deformation twins and stacking faults during plastic deformation at 300K.

Titanium

In early work[352] on nanocrystalline titanium prepared by ball-milling, the relationship between its hardness and the grain-size could be described by,

$$H_v = 2.24(GPa) + k(20MPa\sqrt{mm})d^{-1/2}$$

Detailed analysis of the slope suggested however that it was not absolutely straight and could be divided into two stages. It is moot whether this could have been due to the manifestation of an incipient inverse-Hall-Petch effect. Hydrostatic extrusion[353] of commercial-purity material, in 20 consecutive passes to a cumulative true strain of 5.47, led to grain refinement from about 33μm to 47nm. This refinement was associated with a marked increase in the mechanical properties, while the ductility remained at about 8%. The ultimate tensile strength was greater than 1300MPa and the yield stress was about 1250MPa. The Hall-Petch plots of strength and hardness were not equivalent: the yield-stress dependence was best fitted by a straight line whereas the hardness dependence required a slope change from positive to negative, with the change occurring at about 120nm. This was the point which was commonly assumed to mark where a microstructure changed from being ultra fine-grained to being nanocrystalline. In the present context, it marked the onset of inverse-Hall-Petch behaviour.

Molecular dynamics simulation was used to model[354] the tensile behavior of nanocrystalline titanium, having a grain-size between 2.8 and 10.2nm, at strain-rates ranging from 10^8 to 10^{10}/s. The three-dimensional samples were based upon Voronoi tessellations involving randomly-oriented grains with no texture. Yielding was found always to be controlled only by grain-boundary processes, and an inverse-Hall-Petch relationship between grain size and flow stress was found. The strain-rate sensitivity increased with decreasing grain-size, due to an increase in grain-boundary controlled processes. It also increased with applied strain-rate due to local disorder at the grain boundaries. Marked grain coarsening, in the case of 2.8nm samples, caused a slight increase in the flow stress. Grain-boundary controlled processes and partial dislocation slip both played an important part in the plastic deformation. With increasing grain-size, a twin occasionally initiated from a grain boundary. A size-dependent dislocation-density analysis indicated that, with increasing grain-size, dislocation-related deformation made a greater contribution to plastic strain.

Dense nanocrystalline compacts of Ti-47Al-3at%Cr were produced[355] by mechanical alloying and hot isostatic pressing at 725C, and annealed for up to 800h at 725 to 1200C. Grain-growth occurred, with an exponent which decreased from 10 to 4.6 as the temperature increased. The grain-growth kinetics could be attributed to a single thermally-activated process which was limited by a permanent pinning effect upon the grain boundaries. The microhardness decreased with annealing and could be described by a Hall-Petch equation of the form,

$$H_V = 5.8(GPa) + 1.6(MPa\sqrt{m})d^{-1/2}$$

Nanoparticles of Ti_2B and TiB_2, 5nm in size, were introduced into the grain boundaries of nanocrystalline $Ti_{50}Ni_{25}Cu_{25}$ alloy, increasing the microhardness by 20% and thus closely approaching the theoretical limit[356,357]. The borides tended to suppress low-temperature grain-boundary sliding and thereby shifted the range of anomalous Hall-Petch behaviour towards smaller nanocrystal sizes.

Zirconium

The behavior of nanocrystalline thin films was investigated[358] *in situ* within a transmission electron microscope. The grain-size could then be controlled by electromigration stressing and temperature. The yield stress of specimens having a grain-size of less than 10nm was of the order of 450 to 500MPa, as compared with levels of 250 to 300MPa for bulk material. The fracture stress was about 0.9GPa and the strain-to-fracture some 1.5 to 2%. The results suggested that the critical grain-size for the onset of an inverse-Hall-Petch effect in nanocrystalline hexagonal close-packed metals was about 15nm.

Molecular dynamics methods were used[359] to model the tensile deformation of columnar-grained samples of nanocrystalline zirconium having random rotations around the [00•1] axis. When below a critical grain-size, an inverse-Hall-Petch effect was predicted to occur. At larger grain-sizes, the calculated Hall-Petch constant agreed well with experimental data.

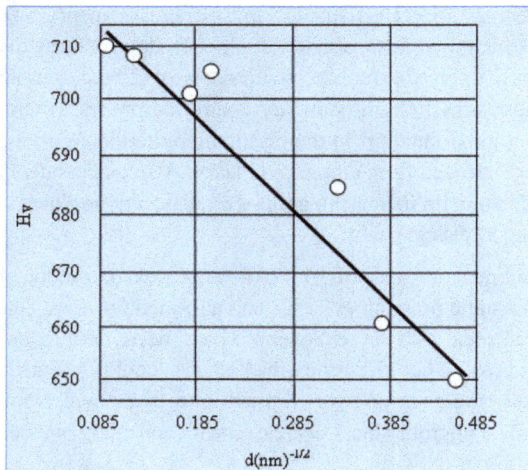

Figure 37. Microhardness of nanocrystalline $Zr_{67}(Fe,Co)_{33}$

The composition, $Zr_{67}(Fe,Co)_{33}$, serves as a potentially interesting exception to the normal Hall-Petch law in that the inverse behaviour tends habitually to be the more common one (figure 37). The nanocrystalline material was here prepared[360] by the crystallization of amorphous precursors, and the hardness was measured over a wide range of grain sizes. The inverse behaviour which observed at extremely small grain sizes was attributed to the presence of remnants of amorphous material. Fully crystalline material was found only at grain-sizes greater than about 20nm but, even there, the hardness did not obey the normal Hall-Petch law. No dislocations were observed in the samples, thus suggesting that deformation was due entirely to grain-boundary sliding.

Thin films of $Zr_{1-x}Mo_x$, with x ranging from 0.32 to 0.95, were deposited[361] onto glass substrates by co-sputtering molybdenum and zirconium. There was a change in the samples, from a nanocrystalline solid-solution of zirconium in the body-centred cubic molybdenum lattice, to clusters of nanocrystalline Zr(Mo) in an amorphous matrix. The coherence length ranged from 1 to 8nm, depending upon the composition. The films had a low coefficient of friction, a high hardness, low Young's modulus and a consequent high hardness/modulus ratio as compared with those of bulk zirconium or molybdenum. An inverse Hall-Petch effect was observed at coherence lengths of less than 6nm.

Cobalt

Plastic flow in nanocrystalline hexagonal close-packed metal, having a grain-size of 4 to 12nm, under uniaxial tensile deformation was modelled[362] by using molecular dynamics simulation and a semi-empirical tight-binding potential. Under stresses of 1.2 to 3.2GPa, at temperatures of 77 to 470K, the growth of disordered atom-segments and their interaction with stacking-faults dominated deformation; even when the grain-size was as small as 4nm. The strain-rate and inverse-Hall-Petch behaviour at grain-sizes of less than 10nm could be accurately described by a disordered atom-segment model for plastic flow.

Returning finally to body-centred cubic materials.

Manganese

Nanocrystalline α-manganese samples having a mean grain-size of 3nm were prepared[363] by electrodeposition, and samples with various grain-sizes were then obtained by annealing the nanocrystalline material in vacuum at various temperatures. The samples exhibited differing Hall-Petch behaviours, ranging from normal to inverse, with decreasing grain-size; the change occurring at a critical grain-size of 42nm.

Niobium

Nanocrystalline powders of Nb-23at%Al were prepared[364] by mechanical alloying and then subjected to pressures of up to 450MPa so as to produce a green density having between 75 and 85% of the theoretical density. The powders were then consolidated to a relative density of up to 99% by high-voltage high current-density pulsed discharge. The consolidated material was a mixture of nanocrystalline Nb_2Al and Nb_3Al, with the grain-size ranging from 13 to 33nm. There was a negative Hall-Petch relationship between the Vickers microhardness and the grain-size, clearly indicating that grain-size softening had occurred.

Tantalum

It was shown[365] that this highly ductile metal could exhibit shear-localization at low temperatures during high strain-rate deformation. The temperature and strain-rate sensitivity for twinning were lower than for slip but, on the other hand, the Hall-Petch slope was higher. The strain-rate therefore determined the predominant deformation mechanism. Using a constitutive equation it was possible to identify a twinning-domain in a Weertman-Ashby plot. A previously-developed constitutive description which incorporated the grain-size dependence of the yield stress was extended into the nanocrystalline range. Simulations permitted the prediction of work-hardening as a function of the grain-size, and the response was successfully modelled for the 50nm to 100μm range. A mechanism for dislocation-generation was presented and provided a constitutive description of plastic deformation. The calculated dislocation densities agreed with observations. Model polycrystals having grain-sizes ranging from 2.5 to 30nm were generated[366] by Voronoi tessellation and subjected to uniaxial tension or compression at strain-rates of between 10^8 and 10^9/s using molecular dynamics simulations. Under tension, fracture began at grain boundaries which were perpendicular to the loading direction and propagated along grain boundaries - with limited plastic deformation – under a stress ranging from 10 to 14GPa. This brittle intergranular failure was attributed to the high strain-rate which was imposed by molecular dynamics; thus leading to a stress which exceeded the grain-boundary cohesive strength. Grain-boundary separation was therefore the main failure mechanism. During compression, considerable plastic deformation occurred within the grains at stresses which were higher than those for failure in tension. The difference between the tensile and compressive failure responses for tantalum was attributed to a difficulty in generating dislocations. The compressive yield stress increased with grain-size, between 2.5 and 30nm. This inverse-Hall-Petch behaviour was analyzed in terms of the contributions which dislocation-motion and grain-boundary shear made to the plastic deformation. As the grain-size increased, the contribution made by grain-boundary sliding decreased and the plastic strain was accommodated by dislocation motion. This behaviour was not observed during tensile deformation. The effect of grain-size upon the flow stress, and on the deformation mechanism of nanocrystalline material under uniaxial tension, was further investigated[367] by using molecular dynamics simulations. This showed that there was a critical grain-size of 7nm at which the flow stress was a maximum. Stacking-fault energy curves indicated that <111>{110} and <111>{112} were the slip systems on which dislocations and twins were most likely to appear, followed by the activation of other slip systems. In samples having a grain-size greater than 7nm, the variation of the flow stress as a function of grain-size obeyed the Hall-Petch relationship and this was attributed to a strengthening effect due to the accumulation of dislocations and twins with decreasing grain-size. In samples having a grain-size of less than 7nm, the variation of flow stress as a function of grain-size exhibited an inverse-Hall-Petch relationship. This was attributed to a softening which was produced by grain-boundary activity. Cracks were found in samples having larger grain-sizes, but these did not propagate to an appreciable extent and thus did not affect the flow stress.

The effect of strong shock waves upon nanocrystalline tantalum was simulated[368] by using atomistic molecular dynamics methods. The particle velocities ranged from 0.35 to 2.0km/s, and induced pressures of 20 to 195GPa. The strain-rates ranged from 10^8 to 10^{10}/s, and led to peak strengths of 3 to 15GPa. The resultant deformation was enabled by dislocation, twinning and/or grain-boundary activity at the various particle velocities. Twinning was observed at a grain-size of 7nm; starting at about 32GPa. This result disagreed with models which predicted Hall-Petch behavior for twinning, in that there was an inverse-Hall-Petch relationship for twinning at small grain-sizes.

Molecular dynamics simulations[369] of the nano-indentation of monocrystalline nano-twinned films showed that the inelastic deformation of monocrystalline tantalum by indentation could be largely attributed to the propagation of dislocation loops and the emergence of deformation twins. An inverse-Hall-Petch relationship applied to the various twin thicknesses. Dislocation absorption, desorption and direct transmission were three dislocation and twin-boundary interactions which resulted in the migration of twin boundaries, the formation of steps along twin boundaries and the formation of new sites for dislocation nucleation; all of which contributed to the softening effect. But the twin boundaries could also block the propagation and migration of dislocations, and thus limit dislocations to the twin lamellae; thereby promoting hardening. Softening mechanisms dominated however the deformation of nano-twinned tantalum during nano-indentation at small twin thicknesses.

Tungsten

Molecular dynamics simulation was used[370] to identify various aspects of the deformation of polycrystalline tungsten nanowire by using the EAM potential and a constant strain-rate of 10^9/s. The average grain-size was varied from 4.63 to 25nm while maintaining a constant diameter of 5nm at a temperature of 10K, or the diameter was varied from 2 to 5nm while keeping the average grain-size and length/diameter ratio constant or the temperature was varied from 10 to 500K while maintaining a fixed diameter and grain-size. Nanowires with a smaller diameter were found to be stronger while high temperatures weakened the nanowires, regardless of the grain-size. The inverse-Hall-Petch law dominated the plasticity of the polycrystalline nanowires and grain-boundary sliding became the predominant plasticity mechanism at a critical grain-size.

Iron

High-energy ball-milling of commercial-purity iron powder for 100h produced[371] nanocrystalline material with a grain-size of 20 to 50nm which was work-hardened to $950H_v$. The hardness of the milled powder depended mainly upon the crystallite size, dislocation cell-size, sub-grains and nanocrystalline grains. The hardness obeyed the Hall-Petch law, with a strain-hardening of $500H_v$. The work-hardening mechanism gradually changed from strain-hardening to grain-refining strengthening at about $600H_v$, the point where the grain-size had been reduced to less than 100nm by heavy deformation. When the plastic deformation of nanocrystalline material was investigated by nano-indentation, a negative Hall-Petch slope was found[372] for volume-averaged grain-sizes which were less than 18nm. The strain-rate sensitivity increased with decreasing grain-size, due to grain-boundary sliding at room temperature.

In an early attempt[373] to explain the inverse-Hall-Petch effect, it was pointed out that, in conventional polycrystalline materials, grain boundaries are considered to be a strengthening factor because they act as barriers to dislocation motion. If the shear displacement that is associated with the impinging slip-bands is to be accommodated plastically in the adjacent grain, the stress-concentration at the grain boundary must be high enough to initiate slip in the neighbouring grain. Refinement of the grains thus results in more grain boundaries and, for a given grain-boundary structure, a higher applied stress should be needed in order to create a dislocation pile-up which is sufficient to make the boundary yield. The sample is thus strengthened. The stress required to cause this also depends however upon the grain-boundary structure and its energetic state. In the case of a high-angle grain boundary, for which the excess volume and excess energy are relatively large, the force required to initiate successive dislocation motions in neighbouring grains would be larger than that required in the case of a small-angle

Materials Research Forum LLC
https://doi.org/10.21741/9781644900352

boundary, where the interfacial energy and interfacial excess volume are smaller. In nanocrystalline samples, a reduction in grain-size results not only in an increase in the number of grain boundaries, and interface volume fraction, but also in a decrease in the interfacial energy and interfacial excess volume. A decrease in the latter would then reduce the ability of the boundary to obstruct dislocation motion. The yield stress therefore decreases and the material is softened. The overall conclusion is that a conventional grain boundary can be a good obstacle to a dislocation pile-up within one grain when the applied stress is smaller than the yield stress, but a nanocrystalline grain-boundary is unable to halt dislocation motion and this may extend into several neighbouring grains. This weakens the material. Some years later it became possible to explore these concepts more quantitatively. Molecular dynamics simulation was used[374] to model the effect of grain-size upon the deformation of nanocrystalline body-centered cubic iron. The average flow stress decreased for grain-sizes of less than 14.7nm, in line with the expected inverse-Hall-Petch effect. This was traced to a change, in the predominant deformation mechanism, from dislocation-glide to grain-boundary sliding. In a similar vein, the tensile deformation of bulk nanocrystalline material was modelled[375] by means of molecular dynamics simulation. The average flow stress here decreased for grain-sizes of less than 13.54nm, again reproducing the breakdown of the normal Hall-Petch law. Again the result was attributed to a change from grain-boundary controlled dislocation activity to grain-boundary phenomena. The results here also showed that the average flow stress should increase with increasing strain-rate and decreasing temperature. Stress-induced phase transformations occurred moreover during tensile deformation and were reversible with regard to the applied stress. The maximum fraction of close-packed atoms also increased with increasing applied stress. Considerable phase transformation, due to dislocation activity, occurred in the stacking-fault zone at a grain-size of 13.54nm, whereas appreciable phase transformation, due to grain-boundary activity, occurred in the grain boundaries at a grain-size of 3.39nm. At deformation temperatures above 900K, essentially no phase transformation occurred because all of the atoms, whether located in the grain boundaries or their interiors, could easily rearrange themselves under thermal activation and form local vacancies or disordered structures, rather than ordered close-packed structures. The effects of grain-size and temperature upon the mechanical behavior of nanocrystalline α-iron under uniaxial tensile loading were further studied[376] via molecular dynamics simulation of a relaxed equilibrium three-dimensional Voronoi construction with grain-sizes of 3.95, 6.80, 9.70, 12.50, 15.50, 17.50, 20.70 or 26.00nm. The constructions were strained (1.0fs time-step) in 0.001 increments (up to 0.2) along the z-direction while maintaining zero stresses in the x- and y-directions. The results showed that the peak stress decreased with decreasing grain-size, yet again reproducing the breakdown of the normal Hall-Petch law, when the grain size was smaller than a critical value. The main deformation mechanism changed, from dislocation-slip and twinning-mediated plasticity at large grain-sizes, to grain-boundary sliding at small grain-sizes. Twinning occurred via the emission of $\frac{1}{6}<111>$ partial dislocations on the {112} slip-plane. Crack formation was the cause of a reduced flow stress in material with a large grain-size, and the Young's modulus decreased with decreasing grain-size. The main sources of cracks were those grain boundaries which were perpendicular to the loading direction, because they were subjected to higher stresses and twin-bands interacted with grain boundaries at larger grain-sizes, thus causing the stress to concentrate at the intersections of grain boundaries. The simulations also exhibited de-twinning, and the migration of deformed twins. De-twinning occurred via migration of the intersection of a grain boundary and a twin. Migration of the deformed twins involved the repeated nucleation and glide of $\frac{1}{6}<111>$ partial dislocations on adjacent {112} planes. The nucleation and propagation of dislocation became easier at higher temperatures. In the very latest molecular dynamics simulations[377], attention was paid to those models which treat nanocrystals as composites in which the grains and grain boundaries behave as two separate phases with their own yield behavior. Estimates of the grain-boundary yield stress were made by modelling indentation. Nanocrystalline iron with a Σ5-type boundary was indented in the grain interior and at the grain boundary. This showed that the grain boundaries yielded at less than half the stress required for grains. This largely confirmed the suspected explanation of the inverse-Hall-Petch effect because, when the grain-size decreases below a critical value, the grain boundary thickness is not negligible and its low yield stress dominates the behaviour of the material. In order to stabilize the boundary with respect to migration or sliding, and to incorporate the influence of segregation, the simulations were repeated but with the addition of 0.008%C as an impurity. This very low carbon content, neatly illustrating the sensitivity of grain boundaries to segregant atoms, nevertheless resulted in a higher grain-boundary yield stress without however increasing the yield stress of the grain interior.

Molecular dynamics simulations were made[378] of the effect of twin-boundary spacing, and of the angle between the load-axis and twin boundary, upon the behaviour of nano-twinned specimens during uniaxial tensile testing. Deformation twins are observed in monocrystalline iron, and are prominent in plastic deformation. The elastic modulus of nano-twinned iron increased slightly with increasing twin-boundary spacing. When the tensile load was not perpendicular to the twin boundary, the yield stress decreased and the deformation involved mainly de-twinning. The simulations indicated that the yield stress of nano-twinned iron increased with increasing twin-boundary spacing; an inverse-Hall-Petch relationship..

Molecular dynamics simulations were made[379] of the plastic deformation of gradient-nanograined metals with regard to the inverse-Hall-Petch relationship, the normal Hall-Petch relationship and their combination. These predicted that the yield strength of gradient-nanograined iron would exceed that of random-nanograined iron; in good agreement with experimental observations. This was due to the fact that strong gradients of stress and strain occur during the deformation of gradient-nanograined structures. The deformation of the middle region of a nanograined film depended upon the stress gradient, and deformation of the surface layer was controlled by the strain gradient. The model showed that the plastic strain gradient depended upon the distribution and size of the nanoscale grains; especially in the case of those below the critical grain-size.

Bulk, powder or ball-milled samples of IF steel (99.96%Fe), Armco steel (99.94%Fe), and of other alloys containing 99.88 or 97.78%Fe, were subjected[380] to high-pressure torsion. The steady-state hardness and tensile strengths of specimens having micron or sub-micron grain-sizes tended to obey the same Hall-Petch relationship as that for pure iron and mild steel. Nano-grained specimens instead followed an inverse-Hall-Petch relationship. The softening in the inverse-Hall-Petch effect could be largely avoided by stabilizing the grain boundaries using carbon atoms, thus showing that any extra hardening caused by impurity additions is due mainly to a grain-boundary strengthening mechanism.

Examination[381] of the microstructure and room-temperature strength of mechanically-alloyed oxide-dispersion strengthened Fe-40Al alloy showed that sub-micron grained materials containing small oxide particles exhibited appreciable hardening due to the particles, plus a similarly significant hardening which was due to the small grain-size. The texture which was created during processing affected the strengthening by governing the barrier efficiency of low-angle or high-angle grain boundaries, thus modifying the Hall-Petch slope for grain-size strengthening. High-energy mechanical milling and cold-pressing was used[382] to produce bulk nanocrystalline FeAl with a grain-size of 40nm. Annealing at up to 480C for 2h permitted the supersaturated Fe(Al) solid solution to precipitate fine metastable Al_5Fe_2, $Al_{13}Fe_4$ and Fe_3Al. Low-temperature annealing relaxed the disordered structure by removing the defects which were introduced by the severe plastic deformation. Peak hardening was found after isochronal aging at 480C. The microhardness increased with grain-size reduction, in accord with the Hall-Petch law, down to a grain-size of 74nm but then decreased at smaller grain sizes.

Bulk nanocrystalline Fe_3Al containing 10%Ni was prepared[383] by aluminothermic reaction and annealed at 800C for various times. The average grain-size fluctuated for annealing times of up to 24h and then became stable. The hardness attained a maximum of $490H_V$ after annealing for 16h, and a minimum of $410H_V$ after annealing for 24h. The smallest grain-size (16nm) was obtained by annealing for 4h and the maximum (35nm) was obtained by annealing for 24h. A critical grain-size (20nm) was identified which separated normal and inverse Hall-Petch behaviours.

High-energy ball-milling was used[384] to prepare monophase nanocrystalline powders of $Fe_{85}Al_4Si_{11}$, $Fe_{90}Ni_{10}$, iron and of nickel alloys containing 1 or 10at%Fe. The grain-size was then increased by annealing. Heating at homologous temperatures of up to 0.4 softened the iron but hardened the body-centered cubic and face-centered cubic solid solutions, where an inverse-Hall-Petch relationship was then observed and attributed to solute segregation to the grain boundaries. Bulk nanocrystalline Fe_3Al containing 10%Mn was prepared by aluminothermic reaction and hot-pressing. The grain-size decreased with increasing hot-pressing time at a given temperature, and increased with increasing hot-pressing temperature for a given time. The hardness decreased with increasing hot-pressing time at 800C, and increased with increasing hot-pressing temperature. The variation of the hardness as a function of grain-size was contrary to the normal Hall-Petch law[385].

Nanocrystalline $Fe_{78}B_{13}Si_9$ samples which were prepared by crystallization exhibited[386], because of large numbers of defects in the grain boundaries, higher values of microhardness than those of coarse-grained crystalline or amorphous samples having the same composition. The hardness of the nanocrystalline material increased with decreasing grain-size. This normal Hall-Petch relationship was of the form,

$$H_v(GPa) = 2.8 + 47.6d(nm)^{-1/2}$$

A complicated behavior was observed[387] in the nanocrystalline material when the grain-size was smaller than 65nm. No sign of dislocations, post-deformation, had been found. This suggested that it was unlikely that dislocation pile-ups were involved. A reduction in the mean hyperfine field was attributed to a decrease in the numbers of nearest-neighbor atoms on grain boundaries, and was held to be responsible for a softening of nanocrystalline samples having an average crystallite size of less than 65nm. An increasing peak in the phonon-mode near to 900/cm indicated a strong electron-atom coupling in the grain boundaries of nanocrystalline material.

The properties of γ (austenitic), α (ferritic) and γ-α duplex stainless steels were determined[388] for samples having grain-sizes of 8 to 1000μm and proportions of γ-phase ranging from 35 to 78%. The hardness and tensile strength of the duplex stainless steel were some 1.5 times greater than those of the austenitic or ferritic stainless steels. The high strength of the duplex steel was attributed to the extensive blocking of slip deformation in the γ- or α-phases and to a large misorientation angle at boundaries between the γ- and α-phases due to their differing lattice structures. The ultimate tensile strength of the duplex steel increased, with increasing proportion of γ-phase, up to 50%. It then decreased with further increases in the amount of γ-phase. The mechanical properties improved with decreasing grain-size of the stainless steels, according to the normal Hall–Petch relationship, but the inverse relationship was found for ferritic stainless steel; especially for grain-sizes of 100 to 1000μm, where the size of $Cr_{23}C_6$ precipitates increased with increasing grain-size.

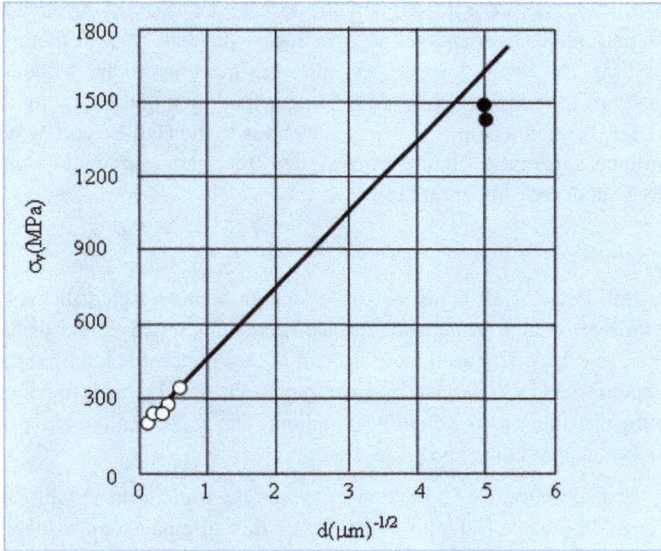

Figure 38. Tensile yield strength of AISI316L as a function of grain-size
Filled circles: present work, open circles: previous data[389]

When Fe-Cr-Ni stainless-steel was prepared by electrodeposition, the nanocrystalline product consisted of α-iron, but this transformed into γ-iron during heat-treatment. Nano-indentation tests showed that the nanocrystalline material was harder than the amorphous material, and this was considered to be consistent with inverse-Hall-Petch behavior[390]. The austenitic stainless steel, 316L, was prepared[391] in nanocrystalline form, with a grain-size of about 40nm, by mechanical attrition. Room-temperature uniaxial tensile tests revealed a yield strength of about 1450MPa, and this value fell on a Hall-Petch plot which was extrapolated from data on coarse-grained material (figure 38).

Nanocrystalline alloy with a grain-size of 15 to 200nm was prepared[392,393] via crystallization of amorphous FeMoSiB. An inverse-Hall-Petch law operated when the average grain-size was below 45nm, while the normal law applied when the grain-size was greater than 45nm. Positron annihilation data indicated that nanovoids in the interfaces markedly affected the microhardness for grain-sizes of less than 45nm.

A study[394] of the melt-quenched amorphous alloys, $Fe_{58}Ni_{25}B_{17}$, $Fe_{50}Ni_{33}B_{17}$ and $Ni_{44}Fe_{29}Co_{15}B_{10}Si_2$, was used to explore the effect of structural parameters upon the strength of amorphous-nanocrystalline specimens. It was shown that, for a given nanoparticle size, the dependence of the microhardness upon the volume fraction or volume density of the nanoparticles could be described by a power-law function with an exponent of 1/3. A relationship which was analogous to the Hall-Petch law was found for a given volume density of nanoparticles, and the hardness could again decrease anomalously with decreasing grain-size.

A Possible Solution

The inverse-Hall-Petch effect is just one of several problems which afflict nanocrystalline metals; the inherently large amounts of surface energy and much shorter diffusion lengths tend to make them thermally unstable. This can in turn impair the long-term strength, but the latter tendency can be combated by encouraging the development of nano-twinning, a bimodal grain distribution or solute segregation. The latter feature, in particular, can inhibit grain-boundary sliding and grain-rotation.

This ability is exemplified by platinum-gold alloys, which can exhibit an incredible resistance to sliding-wear ($10^{-9} mm^3/Nm$)[395] and this originates from the fact that the incidence of boundary motion can be essentially eliminated by incorporating solutes which segregate preferentially at grain boundaries and, most importantly, reduce the grain-boundary free energy. As a result, the destabilising effect of the relatively high grain-boundary surface area can be largely eliminated, thus permitting a material to maintain its microstructure indefinitely at high temperatures. Other methods of impeding grain-growth merely slow the process rather than almost stopping it. Some of the nanocrystalline nickel-tungsten alloys mentioned elsewhere in the present work also exhibit negligible grain growth, during 24h heating at up to 875K, because of tungsten segregation to the grain boundaries and consequent energy-lowering.

As well as promoting perfect thermal stability, the implication is that such solute-segregation manipulation might be exploited in order to remove the inverse-Hall-Petch problem, given that it seems to arise largely from the increased proportion of grain-boundary material present in nanocrystalline alloys. If the mechanical processes occurring in that proportion of the microstructure can be similarly influenced by solute selection, the problematic softening might be avoided. On the other hand, there might have to be a compromise between solving the inverse-Hall-Petch problem and achieving thermal stability. It is already recognized that the tailoring of alloys with regard to thermal stability can impair mechanical properties, with grain-boundary segregating elements in nanocrystalline metals being capable of both strengthening or embrittling those boundaries.

One theory proposes that a solute will embrittle when the solute-host pairing has a positive heat-of-mixing and when the solute possesses a lower surface energy than that of the host. Pairings which involve either a higher heat-of-mixing or a lower solute/solvent ratio of surface energies are also expected to be more embrittling. In the case of the above-mentioned Pt-Au alloy, and its diamond-like wear properties, it has already been confirmed[396] that the solute segregation has an embrittling effect. On the other hand, it was also shown that inhomogeneous solute segregation to grain boundaries can lead to a novel toughening mechanism: so-called compositional crack arrest. This helps to dissipate energy by forming nano-crack networks in which cracks are halted at parts of the grain boundary that are impoverished in the embrittling element. That mechanism, together with triple-junction crack-arrest, then simultaneously optimizes both thermal stability and energy dissipation. The somewhat economically impractical Pt–10at%Au alloy was chosen as a test-bed mainly in order to avoid any complications due to oxidation. This alloy was already known to exhibit marked segregation of gold to platinum grain boundaries, and to combine thermal stability with wear resistance. The strength of the alloy was known to be due to a combination of the normal Hall–Petch effect, and other mechanisms such as solid-solution strengthening.

As already noted, the most important result was that grain boundary embrittlement could be made potentially useful by paradoxically turning nano-cracking into a toughening mechanism. The grain-boundary triple-junctions and heterogeneous segregation could arrest nano-cracks. The platinum and Pt–10Au specimens which were studied were columnar, with an average grain-width of some 50nm, and the grain interiors were about 8at%Au in content while the concentrations at the boundaries ranged from 8 to 30at%Au. The Pt-10Au had a 16%-higher (145GPa) Young's modulus than that (125GPa) of platinum. The nanocrystalline platinum exhibited a plastic deformation behaviour which involved a combination of grain growth and transformation from columnar to equiaxed grains. The Pt-10Au alloy exhibited a combination of plastic behavior, due to microstructural changes, and brittle behavior due to intergranular and transgranular cracking. It also exhibited a lower ductility than that of pure platinum. The platinum exhibited a ductile behavior which was dominated by grain growth, while the alloy exhibited little plastic deformation and deformed mainly via the formation of nano-crack networks with little grain growth, reflecting the thermal stability of the microstructure. It finally failed due to the coalescence of intergranular cracks.

Grain-boundary migration, especially of boundaries having a low gold concentration, led to the consumption of smaller grains so as to form larger grains. An increasing gold concentration was associated with decreasing dislocation activity. Dislocation activity was the highest in platinum. Upon adding 5at%Au, dislocation activity was only slightly reduced while boundary migration was largely suppressed. Further increasing of the solute concentration to 10at%Au produced a large decrease in dislocation activity, and samples fractured before any appreciable dislocation activity could occur. An increasing gold concentration increased the tendency to intergranular cracking. In platinum, no stable nano-cracking was observed at strains of 17%. Intergranular cracks formed in Pt-5Au at gold-rich boundaries, at strains as low as 2%. These cracks arrested, and failure due to coalescence occurred at strains of up to 12%. A network of intergranular cracks formed in Pt-10Au at a strain of about 2%, but coalesced at a strain of 4%.

The ratio of grain-boundary area to sample volume is about $1/\mu m$ for microcrystalline metal and about $10^6/\mu m$ for a nanocrystalline metal; assuming 5nm grains.

The elastic modulus is a factor which affects the overall toughness because the recoverable contribution to toughness in the elastic regime, the so-called resilience, depends upon both the modulus and the strength. For a given yield strength, the resilience decreases with increasing modulus increases and impairs the toughness. The modulus tended here to increase with the gold concentration, to the extent of 16%, although a rule-of-mixtures argument suggested that 10% of added gold should have decreased the modulus by 9%. This was taken to reflect the marked effect of grain-boundary bonding upon the elastic behavior of nanocrystalline metals.

An increase in the resistance to yielding was observed in Pt-10Au alloy, as compared to that of pure platinum. The strengthening behavior was attributed to a solid-solution strengthening plus grain-boundary stabilization. It was noted that traditional solid-solution strengthening could be expected because the atomic radius of gold is 166pm, and is thus some 5% smaller than the 175pm radius of the platinum atom; but systems exhibiting greater differences in atomic radius would lead to superior solid-solution strengthening.

In addition to the traditional solid-solution strengthening, nanocrystalline material can be expected to enjoy in addition a solid-solution strengthening which is due to changes, in the shear modulus and Burgers vector, which arise from the fact that solutes affect the flow stresses required for boundary-controlled plasticity at the nanoscale. The degree of strengthening due to this effect increases with decreasing grain size and can be anticipated to contribute strength-increases which are of the order of some tens of mega-Pascals. This strengthening has been attributed to the inhibition of grain-boundary sliding, or grain rotation, by solute atoms in the grain boundaries. Although the gold content of the present alloys provided stability with respect to diffusional and mechanical grain-boundary movement, the two effects were not assumed to be linked mechanism-wise.

As noted above, the gold additions led moreover to a reduced ductility, with platinum exhibiting more extensive necking as compared with Pt-Au alloys, and grain growth was observed to be a deformation mechanism. There was also a tendency to brittle fracture in the case of Pt-Au specimens. Lower concentrations (5%) of gold permitted much greater plastic deformation; up to 12% strain before failure. This was attributed to a competition between grain-boundary embrittlement and nano-crack toughening.

Control of these competing mechanisms can perhaps be turned into a strategy which permits nanocrystalline metal to retain its toughness even when an embrittling element is added. In the present case, the Pt–Au alloy is expected to be somewhat embrittled because it has a positive heat of mixing of 2.27 and a surface-energy ratio of 0.59. The embrittled material nevertheless exhibited some plasticity due to a combination of mechanisms including dislocation motion, grain-boundary migration and nanocracking. In the case of Pt-10%Au, the boundaries tended to be gold-rich, and to be uniquely associated with intergranular crack formation; thus suggesting a weaker bonding of boundaries which are rich in gold as compared with the grain interior. This further implies that an absence of cracking along boundaries which are impoverished in gold means that those boundaries are more strongly bound than are those which are rich in gold. Pure platinum also exhibits greater plastic deformation, via grain growth and dislocation motion, than do Pt–Au alloys.

Intergranular cracking was a common feature of all of the Pt–Au tests, but embrittlement did not immediately cause failure of any of the specimens; crack arrest being observed. The Pt–Au samples exhibited plastic accommodation due to the formation of a nanocrack network. The similar mechanism of microcrack toughening, is frequently observed in brittle materials, where a network of cracks forms but does not promote immediate failure. The cracks accommodate some of the tensile strain and alleviate the stress on neighboring cracks, thus leading to energy dissipation.

In one mechanism, triple-junction crack arrest occurs when a crack which is propagating along a grain boundary intersects a triple junction. The grain boundary, in the present context, must be gold-rich and possess a normal which is essentially parallel to the loading axis. The resultant combination of a maximum separation-stress acting on the boundary, plus weak bonding of the boundary, then leads to brittle fracture. When the crack intersects a triple-junction, it is stopped due to the now-reduced driving force for intergranular propagation along the inclined gold-rich boundary and the higher resistance to transgranular propagation existing in the tougher grain. The fine scale of nanocrystalline metals offers correspondingly more sites where such arrest can occur, and thus leads to finer nanocracking.

On the other hand, so-called compositional crack arrest occurs when a high gold-content boundary intersects a boundary which is lacking in gold. As above, the boundary which contains the incoming crack must be gold-rich and possess a normal which is almost parallel to the loading axis in order to drive crack formation. The crack then stops at the boundary with its low or zero gold concentration because of its greater cohesion. This mechanism depends upon the existence of a compositional gradient along the grain boundaries, and further requires an heterogeneous solute segregation at boundaries; as in the present system. The mechanism depends upon the boundaries being partially gold-saturated, and it is not certain that such a situation can exist.

This suggests that the greatest toughening of Pt–Au alloys can be obtained by combining gold-rich boundaries, for the purpose of crack nucleation, with gold-depleted boundaries which ensure crack arrest. Intermediate gold concentrations also permit more mechanical grain-boundary migration and dislocation activity. The conclusion drawn is that the toughness of this material can be tailored by choosing the gold content, where the overall toughness is due to the combined effect of an embrittlement which decreases the toughness, an energy-dissipation via nanocrack-network formation and a strengthening which is due to solid-solution strengthening and grain-boundary stabilization which can increase the toughness. Changes in the modulus can meanwhile increase or decrease the toughness.

When the loss in toughness arising from embrittlement is minor when compared with the gains arising from strengthening and nanocrack-toughening, the metal can exhibit an increase in overall toughness in the presence of small additions of the embrittling element. Compositional crack arrest has been shown to be more effective than triple-junction crack arrest, and so a maximum in toughness should occur when compositional crack arrest predominates. When nanocrack-toughening and toughening due to strengthening overcome the initial loss in toughness due to embrittlement, it is theoretically possible to increase the strength, in spite of the embrittling element, because of the potential toughening arising from compositional crack arrest.

Materials Research Forum LLC
https://doi.org/10.21741/9781644900352

In contrast, the achievement of ultra-low wear rates in metals requires the control of different mechanisms to those linked to toughness. Consideration has to be given to the deformation which results from cyclic stresses and leads to microstructural changes, defect nucleation and even to the generation of wear particles. Hall-Petch strengthening has always been linked to an increase in the wear resistance of many types of material. In the case of nanocrystalline metals, the applied surface stresses often do not exceed the flow stress in sliding contact and the wear process is dominated by fatigue mechanisms. This in turn depends upon crack formation due to repetitive cyclic contact stresses, and the formation of particles via cohesive failure in a so-called delamination-wear stage.

Coarse-grained specimens exhibit a dislocation-controlled plasticity, cell formation and grain refinement under cyclic loading. Nanocrystalline metals with unstable grain boundaries can exhibit a grain growth which aids cracking. Improving the thermomechanical stability of nanocrystalline alloys can limit stress-driven microstructural changes under high stresses and suppress delamination wear. Alloying traditionally stabilizes grain structures by creating a solute-drag which slows grain-boundary motion, or by creating Zener pinning via the formation of second-phase precipitates. Stabilizing nanocrystallinity by alloying is an alternative possibility: preferential solute segregation to grain boundaries can lead to a reduction in the boundary energy and thus in the driving force for grain growth. Complete nullification of the thermodynamic driving force for grain growth is possible, leading to slowing of grain growth or to thermodynamic stabilization of the nanocrystalline structure. The effect is greatest for alloys which exhibit marked boundary segregation, and causes segregated grain-boundaries to be energetically preferred to solid solutions or second phases. Platinum-gold alloys are here again a good test-bed for studying those thermodynamic properties which are required for ensuring stable nanocrystallinity. The effect of lattice strain upon segregation in this alloy is also minor, with an elastic mismatch of 4% between platinum and gold. A solid-solution of 90% platinum and 10% gold is predicted to possess an unusually high thermal stability in nanocrystalline form and to exhibit ultra-low wear rates. This suggests that the fatigue-driven delamination wear mechanism of nanocrystalline alloys, which leads to wear-rates as low as $10^{-6}mm^3/Nm$, can be largely suppressed. As a result, magnetron-deposited Pt-Au alloys can exhibit wear-rates which resemble that of diamond and out-perform sapphire and Si_3N_4 counter-faces. Such films exhibit columnar structures with large aspect-ratio grains having a diameter of about 40nm. Solute-enrichment occurred at the boundaries and triple-junctions, while solute-depleted zones occurred in grains which were adjacent to the segregated boundaries. The films were also associated with remarkably stable nanocrystalline structures, with no sign of grain-size change even after annealing for up to 168h at up to 500C. The Pt-Au nanocrystalline films were also extremely resistant to grain-growth under cyclic applied shear stresses, and their wear-rates were of the order of 3 x 10^{-9} to 1 x 10^{-9}mm^3/Nm, with friction coefficients of 0.2 to 0.3. Atomic attrition was suggested to be the predominant wear mechanism.

All of these results, albeit for a somewhat economically impractical model alloy, suggest that careful control of the grain-boundary and grain compositions of nanocrystalline metals offers a means for off-setting the inverse-Hall-Petch effect and extrapolating the properties, promised by the normal Hall-Petch law, to the smallest grain-sizes.

References

[1] Hall, E.O., Proceedings of the Physical Society, 64[9] 1951, 747-753. https://doi.org/10.1088/0370-1301/64/9/303

[2] Petch, N.J., Journal of the Iron and Steel Institute, 173, 1953, 25-28.

[3] Gleiter, H., Marquardt, P., Zeitschrift für Metallkunde, 75[4], 1984, 263-267

[4] Li, Y., Bushby, A.J., Dunstan, D.J., Proceedings of the Royal Society A, 472[2190] 2016, 20150890. https://doi.org/10.1098/rspa.2015.0890

[5] Li, Y., Bushby, A.J., Dunstan, D.J., Materialia, 4, 2018, 182-191. https://doi.org/10.1016/j.mtla.2018.08.017

[6] Jang, J.S.C., Koch, C.C., Scripta Metallurgica et Materiala, 24[8] 1990, 1599-1604. https://doi.org/10.1016/0956-716X(90)90439-N

[7] Kimura, Y., Takaki, S., Materials Transactions, JIM, 36[2] 1995, 289-296. https://doi.org/10.2320/matertrans1989.36.289

[8] Rawers, J.C., Korth, G., Nanostructured Materials, 7[1-2] 1996, 25-45. https://doi.org/10.1016/0965-9773(95)00312-6

[9] Rawers, James C., Journal of Materials Synthesis and Processing, 3[1] 1995, 69-79.

[10]   Savader, J.B., Scanlon, M.R., Cammarata, R.C., Smith, D.T., Hayzelden, C., Scripta Materialia, 36[1] 1997, 29-34. https://doi.org/10.1016/S1359-6462(96)00349-1

[11]   Malow, T.R., Koch, C.C., Metallurgical and Materials Transactions A, 29[9] 1998, 2285-2295. https://doi.org/10.1007/s11661-998-0106-1

[12]   Kawamura, S., Kuwano, H., Takeda, Y., Takahashi, S., Kaga, H., Journal of the Japan Society of Powder and Powder Metallurgy, 50[12] 2003, 1052-1056. https://doi.org/10.2497/jjspm.50.1052

[13] Vetterick, G., Leff, A.C., Marshall, M., Baldwin, J.K., Misra, A., Hattar, K., Taheri, M.L., Materials Science and Engineering A, 709, 2018, 339-348. https://doi.org/10.1016/j.msea.2017.09.020

[14] Tanaka, Y., Takaki, S., Tsuchiyama, T., Uemori, R., ISIJ International, 58[10] 2018, 1927-1933. https://doi.org/10.2355/isijinternational.ISIJINT-2018-371

[15] Gao, G., Gao, B., Gui, X., Hu, J., He, J., Tan, Z., Bai, B., Materials Science and Engineering A, 753, 2019, 1-10. https://doi.org/10.1016/j.msea.2019.03.018

[16] Li, Y.J., Kostka, A., Choi, P., Goto, S., Ponge, D., Kirchheim, R., Raabe, D., Acta Materialia, 84, 2015, 110-123. https://doi.org/10.1016/j.actamat.2014.10.027

[17] Deirmina, F., Pellizzari, M., Materials Science and Engineering A, 743, 2019, 349-360. https://doi.org/10.1016/j.msea.2018.11.093

[18] Lei, C., Deng, X., Li, X., Wang, Z., Wang, G., Misra, R.D.K., Journal of Alloys and Compounds, 689, 2016, 718-725. https://doi.org/10.1016/j.jallcom.2016.08.020

[19] Gubicza, J., El-Tahawy, M., Huang, Y., Choi, H., Choe, H., Lábár, J.L., Langdon, T.G., Materials Science and Engineering A, 657, 2016, 215-223. https://doi.org/10.1016/j.msea.2016.01.057

[20] El-Tahawy, M., Huang, Y., Choi, H., Choe, H., Lábár, J.L., Langdon, T.G., Gubicza, J., Materials Science and Engineering A, 682, 2017, 323-331. https://doi.org/10.1016/j.msea.2016.11.066

[21] Yin, F., Cheng, G.J., Xu, R., Zhao, K., Li, Q., Jian, J., Hu, S., Sun, S., An, L., Han, Q., Scripta Materialia, 155, 2018, 26-31. https://doi.org/10.1016/j.scriptamat.2018.06.014

[22] Ba, D.M., Ma, S.N., Meng, F.J., Transactions of Materials and Heat Treatment, 28[5] 2007, 115-119. https://doi.org/10.1016/j.cct.2006.05.001

[23] Liu, X., Zhu, J., Chao, Y., Jiang, J., Wang, J., Acta Physica Sinica, 42[8] 1993, 1272-1277.

[24] Liu, X.D., Ding, B.Z., Hu, Z.Q., Lu, K., Wang, Y.Z., Physica B, 192[4] 1993, 345-350. https://doi.org/10.1016/0921-4526(93)90009-U

[25] Czerwinski, F., Electrochimica Acta, 44[4] 1998, 667-675. https://doi.org/10.1016/S0013-4686(98)00176-5

[26] Nicolenco, A., Tsyntsaru, N., Fornell, J., Pellicer, E., Reklaitis, J., Baltrunas, D., Cesiulis, H., Sort, J., Materials and Design, 139, 2018, 429-438. https://doi.org/10.1016/j.matdes.2017.11.011

[27]   Wu, D., Zhang, J., Huang, J.C., Bei, H., Nieh, T.G., Scripta Materialia, 68[2] 2013, 118-121. https://doi.org/10.1016/j.scriptamat.2012.09.025

[28]   Jankowski, A.F., Hayes, J.P., Saw, C.K., Philosophical Magazine, 87[16] 2007, 2323-2334. https://doi.org/10.1080/14786430601175532

[29]   Remington, T.P., Hahn, E.N., Zhao, S., Flanagan, R., Mertens, J.C.E., Sabbaghianrad, S., Langdon, T.G., Wehrenberg, C.E., Maddox, B.R., Swift, D.C., Remington, B.A., Chawla, N., Meyers, M.A., Acta Materialia, 158, 2018, 313-329. https://doi.org/10.1016/j.actamat.2018.07.048

[30]   Xin, S.W., Zhang, M., Yang, T.T., Zhao, Y.Y., Sun, B.R., Shen, T.D., Journal of Alloys and Compounds, 769, 2018, 597-604. https://doi.org/10.1016/j.jallcom.2018.07.331

[31]   Cho, Y.S., Coch, C.C., Materials Science and Engineering A, 141[1] 1991, 139-148. https://doi.org/10.1016/0921-5093(91)90717-2

[32]   Ma, C., Wang, S.C., Wood, R.J.K., Zekonyte, J., Luo, Q., Walsh, F.C., Metals and Materials International, 19[6] 2013, 1187-1192. https://doi.org/10.1007/s12540-013-6006-y

[33]   Qiao, G.Y., Xiao, F.R., Materials Research Express, 4[8] 2017, 086512. https://doi.org/10.1088/2053-1591/aa84c4

[34]   Bohn, R., Klassen, T., Bormann, R., Acta Materialia, 49[2] 2001, 299-311. https://doi.org/10.1016/S1359-6454(00)00312-8

[35]   Farhang, M.R., Kamali, A.R., Nazarian-Samani, M., Materials Science and Engineering B: Solid-State Materials for Advanced Technology, 168[1] 2010, 136-141. https://doi.org/10.1016/j.mseb.2009.10.032

[36]   Rusakova, A.V., Lubenets, S.V., Fomenko, L.S., Moskalenko, V.A., Fizika Nizkikh Temperatur, 38[10] 2012, 1240-1250.

[37]   Rusakova, A.V., Lubenets, S.V., Fomenko, L.S., Moskalenko, V.A., Low Temperature Physics, 38[10] 2012, 980-988. https://doi.org/10.1063/1.4758787

[38]   Rusakova, A.V., Lubenets, S.V., Fomenko, L.S., Moskalenko, V.A., Smirnov, A.R., Fizika Nizkikh Temperatur, 41[8] 2015, 835-847.

[39]   Moskalenko, V.A., Pohribnaya, Y.M., Smolianets, R.V., Braude, I.S., Low Temperature Physics, 43[12] 2017, 1427-1431. https://doi.org/10.1063/1.5012796

[40]   Moskalenko, V.A., Pohribnaya, Y.M., Smolianets, R.V., Braude, I.S., Fizika Nizkikh Temperatur, 43[12] 2017, 1789-1795. https://doi.org/10.1063/1.5012796

Materials Research Forum LLC
https://doi.org/10.21741/9781644900352

[41]   Bahl, S., Aleti, B.T., Suwas, S., Chatterjee, K., Philosophical Magazine, 98[23] 2018, 2095-2119. https://doi.org/10.1080/14786435.2018.1478141

[42]   Wu, K., Zhang, J.Y., Li, G., Wang, Y.Q., Cui, J.C., Liu, G., Sun, J., Nanotechnology, 28[44] 2017, 445706. https://doi.org/10.1088/1361-6528/aa887f

[43]   Liu, Y.L., Liu, F., Wu, Q., Chen, A.Y., Li, X., Pan, D., Transactions of Nonferrous Metals Society of China, 24[9] 2014, 2870-2876. https://doi.org/10.1016/S1003-6326(14)63420-8

[44]   Bhaskar, P., Dasgupta, A., Sarma, V.S., Mudali, U.K., Saroja, S., Materials Science and Engineering A, 616, 2014, 71-77. https://doi.org/10.1016/j.msea.2014.08.016

[45]   Ding, J., Tian, Y., Wang, L.S., Huang, X., Zheng, H.R., Song, K., Zeng, X.G., Computational Materials Science, 158, 2019, 76-87. https://doi.org/10.1016/j.commatsci.2018.10.019

[46]   Yuan, C., Fu, R., Zhang, F., Zhang, X., Liu, F., Materials Science and Engineering A, 565, 2013, 27-32. https://doi.org/10.1016/j.msea.2012.11.092

[47]   Gao, Z.H., Sun, X.J., Li, J.G., The Chinese Journal of Process Engineering, 8[5] 2008, 1022-1025.

[48]   Mosavat, S.H., Bahrololoom, M.E., Shariat, M.H., Applied Surface Science, 257[20] 2011, 8311-8316. https://doi.org/10.1016/j.apsusc.2011.03.017

[49]   Feng, Z., Li, Q., Zhang, J., Yang, P., Song, H., An, M., Surface and Coatings Technology, 270, 2015, 47-56. https://doi.org/10.1016/j.surfcoat.2015.03.020

[50]   Sakai, S., Tanimoto, H., Mizubayashi, H., Acta Materialia, 47[1] 1998, 211-217. https://doi.org/10.1016/S1359-6454(98)00339-5

[51]   Tanimoto, H., Sakai, S., Mizubayashi, H., Nanostructured Materials, 12[5] 1999, 751-756. https://doi.org/10.1016/S0965-9773(99)00230-5

[52]   Wang, L., Prorok, B.C., Journal of Materials Research, 23[1] 2008, 55-65. https://doi.org/10.1557/JMR.2008.0032

[53]   Wang, L., Prorok, B.C., Materials Science Forum, 633-634, 2010, 99-105. https://doi.org/10.4028/www.scientific.net/MSF.633-634.99

[54]   Wang, J., Sansoz, F., Deng, C., Xu, G., Han, G., Mao, S.X., Nano Letters, 15[6] 2015, 3865-3870. https://doi.org/10.1021/acs.nanolett.5b00694

[55]   Zheng, L.S., Yuan, X.Q., Lei, T., Wang, C.B., Journal of Synthetic Crystals, 46[9] 2017, 1823-1827.

[56]   Wang, Y.M., Jankowski, A.F., Hamza, A.V., Scripta Materialia, 57[4] 2007, 301-304. https://doi.org/10.1016/j.scriptamat.2007.04.046

[57]   Brun, E., Durut, F., Botrel, R., Theobald, M., Legaie, O., Popa, I., Vignal, V., Journal of the Electrochemical Society, 158[4] 2011, D223-D227. https://doi.org/10.1149/1.3554727

[58]   Qin, X.Y., Wu, X.J., Zhang, L.D., Nanostructured Materials, 5[1] 1995, 101-110. https://doi.org/10.1016/0965-9773(95)00003-W

[59]   Kizuka, T., Ichinose, H., Ishida, Y., Journal of Materials Science, 32[6] 1997, 1501-1507. https://doi.org/10.1023/A:1018566303784

[60]   Mineta, T., Saito, T., Yoshihara, T., Sato, H., Materials Science and Engineering A, 754, 2019, 258-264. https://doi.org/10.1016/j.msea.2019.03.101

[61]   Kim, H., Strachan, A., Modelling and Simulation in Materials Science and Engineering, 23[6] 2015, 065012. https://doi.org/10.1088/0965-0393/23/6/065012

[62]   Li, J., Lu, B., Zhou, H., Tian, C., Xian, Y., Hu, G., Xia, R., Physics Letters A, 2019, In press.

[63]   Farhat, Z.N., Ding, Y., Northwood, D.O., Alpas, A.T., Materials Science and Engineering A, 206[2] 1996, 302-313. https://doi.org/10.1016/0921-5093(95)10016-4

[64]   Bonetti, E., Pasquini, L., Sampaolesi, E., Nanostructured Materials, 9[1-8] 1997, 611-614. https://doi.org/10.1016/S0965-9773(97)00137-2

[65]   Haque, M.A., A Saif, M.T., Scripta Materialia, 47[12] 2002, 863-867. https://doi.org/10.1016/S1359-6462(02)00306-8

[66]   Khan, A.S., Suh, Y.S., Chen, X., Takacs, L., Zhang, H., International Journal of Plasticity, 22[2] 2006, 195-209. https://doi.org/10.1016/j.ijplas.2004.07.008

[67]   Choi, H.J., Lee, S.W., Park, J.S., Bae, D.H., Scripta Materialia, 59[10] 2008, 1123-1126. https://doi.org/10.1016/j.scriptamat.2008.07.030

[68]   Choi, H.J., Lee, S.W., Park, J.S., Bae, D.H., Materials Transactions, 50[3] 2009, 640-643. https://doi.org/10.2320/matertrans.MRA2008343

[69]   Ito, Y., Edalati, K., Horita, Z., Materials Science and Engineering A, 679, 2017, 428-434. https://doi.org/10.1016/j.msea.2016.10.066

[70]   Matsui, I., Ono, S., Takigawa, Y., Uesugi, T., Higashi, K., Materials Science and Engineering A, 550, 2012, 363-366. https://doi.org/10.1016/j.msea.2012.04.088

[71]   Tao, J.M., Huang, S.Z., Xu, M.C., Li, C.J., Zhu, X.K., Journal of Materials

Engineering, 6, 2009, 67-72.

[72]   Ferguson, J.B., Tabandeh-Khorshid, M., Rohatgi, P.K., Cho, K., Kim, C.S., Scripta Materialia, 72-73, 2014, 13-16. https://doi.org/10.1016/j.scriptamat.2013.10.005

[73]   Barna, P.B., Adamik, M., Kaiser, U., Laux, S., Bangert, H., Pulliainen, M., Pischow, K.A., Surface and Coatings Technology, 100-101[1-3] 1998, 72-75. https://doi.org/10.1016/S0257-8972(97)00590-2

[74]   Srinivasan, D., Chattopadhyay, K., Materials Science and Engineering A, 375-377[1-2S] 2004, 1228-1234. https://doi.org/10.1016/j.msea.2003.10.189

[75]   Ganji, R.S., Sai Karthik, P., Bhanu Sankara Rao, K., Rajulapati, K.V., Acta Materialia, 125, 2017, 58-68. https://doi.org/10.1016/j.actamat.2016.11.046

[76]   Jin, H., Metallurgical and Materials Transactions A, 49[12] 2018, 6122-6133. https://doi.org/10.1007/s11661-018-4949-9

[77]   Varin, R.A., Wexler, D., Calka, A., Zbroniec, L., Intermetallics, 6[6] 1998, 547-557. https://doi.org/10.1016/S0966-9795(98)00006-5

[78]   Weertman, J.R., Materials Science and Engineering A, 166[1-2] 1993, 161-167. https://doi.org/10.1016/0921-5093(93)90319-A

[79]   Fougere, G.E., Weertman, J.R., Siegel, R.W., Kim, S., Scripta Metallurgica et Materiala, 26[12] 1992, 1879-1883. https://doi.org/10.1016/0956-716X(92)90052-G

[80]   Chokshi, A.H., Rosen, A., Karch, J., Gleiter, H., Scripta Metallurgica, 23[10] 1989, 1679-1683. https://doi.org/10.1016/0036-9748(89)90342-6

[81]   Nieman, G.W., Weertman, J.R., Siegel, R.W., Journal of Materials Research, 6[5] 1991, 1012-1027. https://doi.org/10.1557/JMR.1991.1012

[82]   Sanders, P.G., Eastman, J.A., Weertman, J.R., Acta Materialia, 45[10] 1997, 4019-4025. https://doi.org/10.1016/S1359-6454(97)00092-X

[83]   Youngdahl, C.J., Sanders, P.G., Eastman, J.A., Weertman, J.R., Scripta Materialia, 37[6] 1997, 809-813. https://doi.org/10.1016/S1359-6462(97)00157-7

[84]   Gertsman, V.Y., Valiev, R.Z., Akhmadeev, N.A., Mishin, O.V., Materials Science Forum, 225-227[2] 1996, 739-744. https://doi.org/10.4028/www.scientific.net/MSF.225-227.739

[85]   Gray, G.T., Lowe, T.C., Cady, C.M., Valiev, R.Z., Aleksandrov, I.V., Nanostructured Materials, 9[1-8] 1997, 477-480. https://doi.org/10.1016/S0965-9773(97)00104-9

[86]   Suryanarayanan Iyer, R., Frey, C.A., Sastry, S.M.L., Waller, B.E., Buhro, W.E., Materials Science and Engineering A, 264[1-2] 1999, 210-214. https://doi.org/10.1016/S0921-5093(98)01027-2

[87]   Wang, Y., Chen, M., Zhou, F., Ma, E., Nature, 419[6910] 2002, 912-915. https://doi.org/10.1038/nature01133

[88]   Thilly, L., Lecouturier, F., Von Stebut, J., Acta Materialia, 50[20] 2002, 5049-5065. https://doi.org/10.1016/S1359-6454(02)00351-8

[89]   Chu, G., Tang, Y.J., Liu, W., Yang, T.Z., Transactions of Nonferrous Metals Society of China, 16[4] 2006, 873-877. https://doi.org/10.1016/S1003-6326(06)60343-9

[90]   Chen, J., Lu, L., Lu, K., Scripta Materialia, 54[11] 2006, 1913-1918. https://doi.org/10.1016/j.scriptamat.2006.02.022

[91]   Liao, X.Z., Zhao, Y.H., Srinivasan, S.G., Zhu, Y.T., Valiev, R.Z., Gunderov, D.V., Applied Physics Letters, 84[4] 2004, 592-594. https://doi.org/10.1063/1.1644051

[92]   Lamovec, J., Jović, V., Vorkapić, M., Popović, B., Radojević, V., Aleksić, R., Journal of Mining and Metallurgy B, 47[1] 2011, 53-61. https://doi.org/10.2298/JMMB1101053L

[93]   Song, H.Y., Sun, Y., Computational Materials Science, 104, 2015, 46-51. https://doi.org/10.1016/j.commatsci.2015.03.052

[94]   Okamoto, N.L., Kashioka, D., Hirato, T., Inui, H., International Journal of Plasticity, 56, 2014, 173-183. https://doi.org/10.1016/j.ijplas.2013.12.003

[95]   Zhao, P., Guo, Y., Computational Materials Science, 155, 2018, 431-438. https://doi.org/10.1016/j.commatsci.2018.09.014

[96]   Darling, K.A., Tschopp, M.A., Guduru, R.K., Yin, W.H., Wei, Q., Kecskes, L.J., Acta Materialia, 76, 2014, 168-185. https://doi.org/10.1016/j.actamat.2014.04.074

[97]   Azabou, M., Makhlouf, T., Saurina, J., Escoda, L., Suñol, J.J., Khitouni, M., International Journal of Advanced Manufacturing Technology, 87[1-4] 2016, 981-987. https://doi.org/10.1007/s00170-016-8551-2

[98]   Kapoor, M., Kaub, T., Darling, K.A., Boyce, B.L., Thompson, G.B., Acta Materialia, 126, 2017, 564-575. https://doi.org/10.1016/j.actamat.2016.12.057

[99]   Brink, T., Albe, K., Acta Materialia, 156, 2018, 205-214. https://doi.org/10.1016/j.actamat.2018.06.036

[100] Sinha, S., Komarasamy, M., Thapliyal, S., Gwalani, B., Shukla, S., Darling, K.A.,

Mishra, R.S., Scripta Materialia, 164, 2019, 42-47.
https://doi.org/10.1016/j.scriptamat.2019.01.038

[101] Sikdar, K., Mahata, A., Chakravarty, S., Atwater, M.A., Roy, D., Koch, C.C., Materialia, 5, 2019, 100253. https://doi.org/10.1016/j.mtla.2019.100253

[102] Haouala, S., Segurado, J., LLorca, J., Acta Materialia, 148, 2018, 72-85. https://doi.org/10.1016/j.actamat.2018.01.024

[103] Li, J., Weng, G.J., International Journal of Plasticity, 23[12] 2007, 2115-2133. https://doi.org/10.1016/j.ijplas.2007.03.016

[104] Shaw, L.L., Ortiz, A.L., Villegas, J.C., Scripta Materialia, 58[11] 2008, 951-954. https://doi.org/10.1016/j.scriptamat.2008.01.025

[105] Godon, A., Creus, J., Cohendoz, S., Conforto, E., Feaugas, X., Girault, P., Savall, C., Scripta Materialia, 62[6] 2010, 403-406.
https://doi.org/10.1016/j.scriptamat.2009.11.038

[106] Tanaka, K., Isokawa, Y., Asano, H., Kimachi, H., Journal of the Society of Materials Science of Japan, 59[4] 2010, 315-321. https://doi.org/10.2472/jsms.59.315

[107] Ibáñez, A., Escudero-Cid, R., Ocón, P., Fatás, E., Surface and Coatings Technology, 212, 2012, 94-100. https://doi.org/10.1016/j.surfcoat.2012.09.027

[108] Tanaka, K., Koike, Y., Sano, K., Tanaka, H., Machiya, S., Shobu, T., Kimachi, H., Journal of the Society of Materials Science of Japan, 64[7] 2015, 528-535.
https://doi.org/10.2472/jsms.64.528

[109] Pillai, A.M., Rajendra, A., Sharma, A.K., Transactions of the Institute of Metal Finishing, 90[3] 2012, 143-148. https://doi.org/10.1179/0020296712Z.00000000020

[110] Mamaghani, K.R., Naghib, S.M., International Journal of Electrochemical Science, 12[6] 2017, 5023-5035.

[111] Danilov, F.I., Protsenko, V.S., Kityk, A.A., Shaiderov, D.A., Vasileva, E.A., Kumar, U.P., Kennady, C.J., Protection of Metals and Physical Chemistry of Surfaces, 53[6] 2017, 1131-1138. https://doi.org/10.1134/S2070205118010203

[112] Farkas, D., Curtin, W.A., Materials Science and Engineering A, 412[1-2] 2005, 316-322. https://doi.org/10.1016/j.msea.2005.09.043

[113] Wang, Y.M., Ott, R.T., Van Buuren, T., Willey, T.M., Biener, M.M., Hamza, A.V., Physical Review B, 85[1] 2012, 014101.
https://doi.org/10.1103/PhysRevB.85.014101

[114] Yang, B., Vehoff, H., Zeitschrift für Metallkunde, 95[6] 2004, 499-504.

https://doi.org/10.3139/146.017988

[115] Lei, W.N., Zhu, D., Qu, N.S., Transactions of the Institute of Metal Finishing, 82[5-6] 2004, 141-143. https://doi.org/10.1080/00202967.2004.11871579

[116] Lei, W., Zhu, D., Qu, N., Chinese Journal of Mechanical Engineering, 40[12] 2004, 124-127. https://doi.org/10.3901/JME.2004.12.124

[117] Xuetao, Y., Yu, W., Dongbai, S., Hongying, Y., Surface and Coatings Technology, 202[9] 2008, 1895-1903. https://doi.org/10.1016/j.surfcoat.2007.08.023

[118] Ba, Z.X., Dai, Y.M., Zhang, Z.Z., Zhou, J.Q., Zhang, S.M., Transactions of Materials and Heat Treatment, 31[9] 2010, 137-141.

[119] Seo, B.B., Jahed, Z., Burek, M.J., Tsui, T.Y., Materials Science and Engineering A, 596, 2014, 275-284. https://doi.org/10.1016/j.msea.2013.12.035

[120] Feng, L., Ren, Y.Y., Zhang, Y.H., Wang, S., Li, L., Metals, 9[2] 2019, 188. https://doi.org/10.3390/met9020188

[121] Choudry, M.S., Dollar, M., Eastman, J.A., Materials Science and Engineering A, 256[1-2] 1998, 25-33. https://doi.org/10.1016/S0921-5093(98)00810-7

[122] Chen, T., Hampikian, J.M., Thadhani, N.N., Acta Materialia, 47[8] 1999, 2567-2579. https://doi.org/10.1016/S1359-6454(99)00059-2

[123] Xiao, C., Mirshams, R.A., Whang, S.H., Yin, W.M., Materials Science and Engineering A, 301[1] 2001, 35-43. https://doi.org/10.1016/S0921-5093(00)01392-7

[124] Matsui, I., Uesugi, T., Takigawa, Y., Higashi, K., Acta Materialia, 61[9] 2013, 3360-3369. https://doi.org/10.1016/j.actamat.2013.02.025

[125] Li, Y., Jiang, H., Wang, D., Ge, H., Surface and Coatings Technology, 202[20] 2008, 4952-4956. https://doi.org/10.1016/j.surfcoat.2008.04.093

[126] Bakhit, B., Akbari, A., Journal of Coatings Technology and Research, 10[2] 2013, 285-295. https://doi.org/10.1007/s11998-012-9437-3

[127] Qin, X.Y., Lee, J.S., Lee, C.S., Journal of Materials Research, 17[5] 2002, 991-1001. https://doi.org/10.1557/JMR.2002.0147

[128] Li, H., Ebrahimi, F., Materials Science and Engineering A, 347[1-2] 2003, 93-101. https://doi.org/10.1016/S0921-5093(02)00586-5

[129] Li, H.Q., Ebrahimi, F., Acta Materialia, 51[13] 2003, 3905-3913. https://doi.org/10.1016/S1359-6454(03)00215-5

[130] Ebrahimi, F., Ahmed, Z., Li, H.Q., Materials and Manufacturing Processes, 21[7]

2006, 687-693. https://doi.org/10.1080/10426910600611748

[131] Mulligan, C.P., Wei, R., Yang, G., Zheng, P., Deng, R., Gall, D., Surface and Coatings Technology, 270, 2015, 299-304. https://doi.org/10.1016/j.surfcoat.2015.02.030

[132] Pu, E., Zheng, W., Song, Z., Zhang, K., Yang, F., Lu, H., Dong, H., Materials Science and Engineering A, 705, 2017, 335-347. https://doi.org/10.1016/j.msea.2017.08.101

[133] Li, S., Sun, L., Wang, Z., Nanostructured Materials, 2[6] 1993, 653-661. https://doi.org/10.1016/0965-9773(93)90040-I

[134] Pande, C.S., Masumura, R.A., Armstrong, R.W., Nanostructured Materials, 2[3] 1993, 323-331. https://doi.org/10.1016/0965-9773(93)90159-9

[135] Lian, J., Baudelet, B., Nanostructured Materials, 2[4] 1993, 415-419. https://doi.org/10.1016/0965-9773(93)90184-D

[136] Wang, N., Wang, Z., Aust, K.T., Erb, U., Acta Metallurgica et Materialia, 43[2] 1995, 519-528. https://doi.org/10.1016/0956-7151(94)00253-E

[137] Chokshi, A.H., Scripta Materialia, 34[12] 1996, 1905-1910. https://doi.org/10.1016/1359-6462(96)00071-1

[138] Zaĭchenko, S.G., Glezer, A.M., Physics of the Solid State, 39[11] 1997, 1810-1814. https://doi.org/10.1134/1.1130179

[139] Zaichenko, S.G., Glezer, A.M., Materials Science Forum, 269-272[2] 1998, 687-692. https://doi.org/10.4028/www.scientific.net/MSF.269-272.687

[140] Ovidko, I.A., Nanostructured Materials, 8[2] 1997, 149-153. https://doi.org/10.1016/S0965-9773(97)00006-8

[141] Naimark, O.B., Physics of Metals and Metallography, 84[4] 1997, 327-337.

[142] Sanders, P.G., Youngdahl, C.J., Weertman, J.R., Materials Science and Engineering A, 234-236, 1997, 77-82. https://doi.org/10.1016/S0921-5093(97)00185-8

[143] Huang, G.Y., Peng, X.L., International Journal of Applied Mechanics, 8[4] 2016, 1650046. https://doi.org/10.1142/S1758825116500460

[144] Konstantinidis, D.A., Aifantis, E.C., Nanostructured Materials, 10[7] 1998, 1111-1118. https://doi.org/10.1016/S0965-9773(98)00145-7

[145] Meyers, M.A., Benson, D.J., Fu, H.H., Proceedings of the TMS Fall Meeting,

1999, 499-512.

[146] Fu, H.H., Benson, D.J., Meyers, M.A., Acta Materialia, 49[13] 2001, 2567-2582. https://doi.org/10.1016/S1359-6454(01)00062-3

[147] Benson, D.J., Fu, H.H., Meyers, M.A., Materials Science and Engineering A, 319-321, 2001, 854-861. https://doi.org/10.1016/S0921-5093(00)02029-3

[148] Kim, H.S., Estrin, Y., Bush, M.B., Materials Science and Engineering A, 316[1-2] 2001, 195-199. https://doi.org/10.1016/S0921-5093(01)01246-1

[149] Wen, Y.H., Zhou, F.X., Liu, Y.W., Chinese Physics, 10[5] 2001, 407-412. https://doi.org/10.1088/1009-1963/10/5/309

[150] Jérusalem, A., Philosophical Magazine Letters, 91[9] 2011, 599-609. https://doi.org/10.1080/09500839.2011.598478

[151] Tengen, T.B., Nanotechnology 2010: Technical Proceedings of the 2010 NSTI Nanotechnology Conference and Expo, NSTI-Nanotech., 2, 2010, 661-664.

[152] Lung, C., Tian, E., Ye, H., Journal of Materials Science and Technology, 17[3] 2001, 388.

[153] Tian, D., Zhou, C.J., He, J.H., Fractals, 26[6] 2018, 1850083. https://doi.org/10.1142/S0218348X18500834

[154] Li, J., Yip, S., Computer Modeling in Engineering and Sciences, 3[2] 2002, 219-227.

[155] Capolungo, L., Cherkaoui, M., Qu, J., Transactions of the ASME, Journal of Engineering Materials and Technology, 127[4] 2005, 400-407. https://doi.org/10.1115/1.1925288

[156] Yip, S., Nature Materials, 3[1] 2004, 11-12. https://doi.org/10.1038/nmat1053

[157] Hansen, N., Advanced Engineering Materials, 7[9] 2005, 815-821. https://doi.org/10.1002/adem.200500102

[158] Meyers, M.A., Mishra, A., Benson, D.J., Progress in Materials Science, 51[4] 2006, 427-556. https://doi.org/10.1016/j.pmatsci.2005.08.003

[159] Mohamed, F.A., Metallurgical and Materials Transactions A, 38[2] 2007, 340-347. https://doi.org/10.1007/s11661-006-9057-6

[160] Barai, P., Weng, G.J., Acta Mechanica, 195[1-4] 2008, 327-348. https://doi.org/10.1007/s00707-007-0558-1

[161] Barai, P., Weng, G.J., International Journal of Plasticity, 24[8] 2008, 1380-1410.

https://doi.org/10.1016/j.ijplas.2007.09.010

[162] Weng, G.J., Reviews on Advanced Materials Science, 19[1-2] 2009, 41-62.

[163] Firstov, S.A., Rogul, T.G., Shut, O.A., Materials Science, 45[6] 2009, 759-767. https://doi.org/10.1007/s11003-010-9241-0

[164] Saada, G., Dirrasb, G., International Journal of Materials Research, 100[10] 2009, 1456-1460. https://doi.org/10.3139/146.110197

[165] Winfield, I., Brooks, I., Yokley, E.M., Heard, R., Palumbo, G., Wire Journal International, 42[4] 2009, 102-106.

[166] Shi, J., Zikry, M.A., Materials Science and Engineering A, 520[1-2] 2009, 121-133. https://doi.org/10.1016/j.msea.2009.05.012

[167] Zhang, H., Han, Y., Physical Review X, 8[4] 2018, 041023. https://doi.org/10.1103/PhysRevX.8.041023

[168] Gu, P., Kad, B.K., Dao, M., Scripta Materialia, 62[6] 2010, 361-364. https://doi.org/10.1016/j.scriptamat.2009.10.035

[169] Li, S.X., Ren, X.P., Qu, H.T., Li, S.F., Materials Science and Technology, 30[10] 2014, 1235-1238. https://doi.org/10.1179/1743284713Y.0000000441

[170] Kato, M., Materials Transactions, 55[1] 2014, 19-24. https://doi.org/10.2320/matertrans.MA201310

[171] Quek, S.S., Wu, Z., Zhang, Y.W., Srolovitz, D.J., Acta Materialia, 75, 2014, 92-105. https://doi.org/10.1016/j.actamat.2014.04.063

[172] Pineau, A., Philosophical Transactions of the Royal Society A, 373[2038] 2015, 20140131. https://doi.org/10.1098/rsta.2014.0131

[173] Xiao, X.Z., Song, D.K., Chu, H.J., Xue, J.M., Duan, H.L., Proceedings of the Royal Society A, 471[2177] 2015, 20140832. https://doi.org/10.1098/rspa.2014.0832

[174] Yuan, R., Beyerlein, I.J., Zhou, C., Acta Materialia, 90, 2015, 169-181. https://doi.org/10.1016/j.actamat.2015.02.035

[175] Leitner, A., Maier-Kiener, V., Kiener, D., Advanced Engineering Materials, 19[4] 2017, 1600669. https://doi.org/10.1002/adem.201600669

[176] Abdul-Latif, A., Kerkour-El Miad, A., Baleh, R., Garmestani, H., Mechanics of Materials, 126, 2018, 1-12. https://doi.org/10.1016/j.mechmat.2018.07.002

[177] Borodin, E.N., Mayer, A.E., Materials Research Express, 4[8] 2017, 085040. https://doi.org/10.1088/2053-1591/aa8514

[178] Palumbo, G., Aust, K.T., Erb, U., Materials Science Forum, 225-227[1] 1996, 281-286. https://doi.org/10.4028/www.scientific.net/MSF.225-227.281

[179] Hahn, H., Mondal, P., Padmanabhan, K.A., Nanostructured Materials, 9[1-8] 1997, 603-606. https://doi.org/10.1016/S0965-9773(97)00135-9

[180] Zaichenko, S.G., Glezer, A.M., Interface Science, 7[1] 1999, 56-67. https://doi.org/10.1023/A:1008714612121

[181] Song, H.W., Guo, S.R., Hu, Z.Q., Nanostructured Materials, 11[2] 1999, 203-210. https://doi.org/10.1016/S0965-9773(99)00033-1

[182] Chattopadhyay, P.P., Pabi, S.K., Manna, I., Tomota, Y., Zeitschrift für Metallkunde, 91[12] 2000, 1049-1051.

[183] Narayan, J., Wang, H., Kvit, A., American Society of Mechanical Engineers, 95, 2001, 55-63.

[184] Wolf, D., Yamakov, V., Phillpot, S.R., Mukherjee, A.K., Zeitschrift für Metallkunde, 94[10] 2003, 1091-1097. https://doi.org/10.3139/146.031091

[185] Yamakov, V., Wolf, D., Phillpot, S.R., Mukherjee, A.K., Gleiter, H., Philosophical Magazine Letters, 83[6] 2003, 385-393. https://doi.org/10.1080/09500830031000120891

[186] Tengen, T.B., Wejrzanowski, T., Iwankiewicz, R., Kurzydlowski, K.J., Solid State Phenomena, 140, 2008, 185-190. https://doi.org/10.4028/www.scientific.net/SSP.140.185

[187] Desai, T.G., Millett, P., Wolf, D., Materials Science and Engineering A, 493[1-2] 2008, 41-47. https://doi.org/10.1016/j.msea.2007.06.097

[188] Jiang, B., Weng, G.J., Journal of the Mechanics and Physics of Solids, 52[5] 2004, 1125-1149. https://doi.org/10.1016/j.jmps.2003.09.002

[189] Ramtani, S., Bui, H.Q., Dirras, G., International Journal of Engineering Science, 47[4] 2009, 537-553. https://doi.org/10.1016/j.ijengsci.2008.09.005

[190] Schiotz, J., Materials Science and Engineering A, 375-377[1-2S] 2004, 975-979. https://doi.org/10.1016/j.msea.2003.10.175

[191] Kim, H.S., Estrin, Y., Acta Materialia, 53[3] 2005, 765-772. https://doi.org/10.1016/j.actamat.2004.10.028

[192] Fan, G.J., Choo, H., Liaw, P.K., Lavernia, E.J., Materials Science and Engineering A, 409[1-2] 2005, 243-248. https://doi.org/10.1016/j.msea.2005.06.073

[193] Pande, C.S., Masumura, R.A., Materials Science and Engineering A, 409[1-2] 2005, 125-130. https://doi.org/10.1016/j.msea.2005.04.058

[194] Qing, X., Xingming, G., International Journal of Solids and Structures, 43[25-26] 2006, 7793-7799. https://doi.org/10.1016/j.ijsolstr.2006.04.015

[195] Rodriguez, P., Armstrong, R.W., Bulletin of Materials Science, 29[7] 2006, 717-720.

[196] Carlton, C.E., Ferreira, P.J., Acta Materialia, 55[11] 2007, 3749-3756. https://doi.org/10.1016/j.actamat.2007.02.021

[197] Malygin, G.A., Physics of the Solid State, 50[6] 2008, 1056-1060. https://doi.org/10.1134/S1063783408060103

[198] Zhao, M., Jiang, Q., (2006) NanoSingapore 2006: IEEE Conference on Emerging Technologies - Nanoelectronics - Proceedings, 1609774, 2006, 472-474.

[199] Vo, N.Q., Averback, R.S., Bellon, P., Caro, A., Physical Review B, 78[24] 2008, 241402. https://doi.org/10.1103/PhysRevB.78.241402

[200] Romanov, A.E., Kolesnikova, A.L., Ovidko, I.A., Aifantis, E.C., Materials Science and Engineering A, 503[1-2] 2009, 62-67. https://doi.org/10.1016/j.msea.2008.05.053

[201] Kolesnikova, A.L., Ovidko, I.A., Romanov, A.E., Technical Physics Letters, 33[8] 2007, 641-644. https://doi.org/10.1134/S1063785007080056

[202] Stefanovic, P., Haataja, M., Provatas, N., Physical Review E, 80[4] 2009, 046107. https://doi.org/10.1103/PhysRevE.80.046107

[203] Aifantis, K.E., Konstantinidis, A.A., Materials Science and Engineering B, 163[3] 2009, 139-144. https://doi.org/10.1016/j.mseb.2009.05.010

[204] Gürses, E., El Sayed, T., Materials Letters, 65[23-24] 2011, 3391-3395. https://doi.org/10.1016/j.matlet.2011.07.039

[205] Gürses, E., El Sayed, T., Journal of the Mechanics and Physics of Solids, 59[3] 2011, 732-749. https://doi.org/10.1016/j.jmps.2010.10.010

[206] Li, J., Weng, G.J., International Journal of Mechanics and Materials in Design, 9[2] 2013, 141-152. https://doi.org/10.1007/s10999-013-9214-1

[207] Huang, M., Bouaziz, O., Der Zwaag, S.V., Materials Letters, 65[19-20] 2011, 3128-3130. https://doi.org/10.1016/j.matlet.2011.06.104

[208] Zhang, X., Aifantis, K.E., Journal of Materials Research, 26[11] 2011, 1399-1405.

https://doi.org/10.1557/jmr.2011.123

[209] Belousov, O.K., Palii, N.A., Zabolotnyi, V.T., Russian Metallurgy, 2011[1] 2011, 33-46. https://doi.org/10.1134/S0036029511010046

[210] Zhao, Y., Chen, Z., Long, J., Yang, T., Acta Metallurgica Sinica, 27[1] 2014, 81-86. https://doi.org/10.1007/s40195-014-0027-5

[211] Padmanabhan, K.A., Sripathi, S., Hahn, H., Gleiter, H., Materials Letters, 133, 2014, 151-154. https://doi.org/10.1016/j.matlet.2014.06.153

[212] Hahn, E.N., Meyers, M.A., Materials Science and Engineering A, 646, 2015, 101-134. https://doi.org/10.1016/j.msea.2015.07.075

[213] Quek, S.S., Chooi, Z.H., Wu, Z., Zhang, Y.W., Srolovitz, D.J., Journal of the Mechanics and Physics of Solids, 88, 2016, 252-266. https://doi.org/10.1016/j.jmps.2015.12.012

[214] Tengen, T.B., International Journal of Mechanics and Materials in Design, 8[2] 2012, 101-112. https://doi.org/10.1007/s10999-012-9177-7

[215] Zhao, Y.L., Chen, Z., Long, J., Yang, T., Acta Physica Sinica, 62[11] 2013, 118102.

[216] Dobosz, R., Lewandowska, M., Kurzydlowski, K.J., Scripta Materialia, 67[4] 2012, 408-411. https://doi.org/10.1016/j.scriptamat.2012.05.043

[217] Upadhyay, M.V., Capolungo, L., Taupin, V., Fressengeas, C., Lebensohn, R.A., International Journal of Plasticity, 83, 2016, 126-152. https://doi.org/10.1016/j.ijplas.2016.04.007

[218] Borodin, E.N., Mayer, A.E., Modelling and Simulation in Materials Science and Engineering, 24[2] 2016, 025013. https://doi.org/10.1088/0965-0393/24/2/025013

[219] Klusemann, B., Bargmann, S., Estrin, Y., Modelling and Simulation in Materials Science and Engineering, 24[8] 2016, 085016. https://doi.org/10.1088/0965-0393/24/8/085016

[220] Li, J.J., Xian, Y.H., Zhou, H.J., Wu, R.N., Hu, G.M., Xia, R., Science China Technological Sciences, 61[9] 2018, 1353-1363. https://doi.org/10.1007/s11431-018-9270-9

[221] Tran, A.S., Fang, T.H., Hsiao, J.W., Current Applied Physics, 19[3] 2019, 332-340. https://doi.org/10.1016/j.cap.2018.12.015

[222] Jankowski, A.F., Nyakiti, L.O., Surface and Coatings Technology, 205[5] 2010, 1398-1402. https://doi.org/10.1016/j.surfcoat.2010.07.106

[223] Yuan, L., Jing, P., Liu, Y.H., Xu, Z.H., Shan, D.B., Guo, B., Acta Physica Sinica, 63[1] 2014, 016201.

[224] Nieman, G.W., Weertman, J.R., Siegel, R.W., Scripta Metallurgica et Materiala, 24[1] 1990, 145-150. https://doi.org/10.1016/0956-716X(90)90582-2

[225] Lian, J., Baudelet, B., Nazarov, A.A., Materials Science and Engineering A, 172[1-2] 1993, 23-29. https://doi.org/10.1016/0921-5093(93)90422-B

[226] Ivanisenko, Y., Tabachnikova, E.D., Psaruk, I.A., Smirnov, S.N., Kilmametov, A., Kobler, A., Kübel, C., Kurmanaeva, L., Csach, K., Mishkuf, Y., Scherer, T., Semerenko, Y.A., Hahn, H., International Journal of Plasticity, 60, 2014, 40-57. https://doi.org/10.1016/j.ijplas.2014.04.011

[227] Butt, M.Z., Javed, A., Khaliq, M.W., Ali, D., Bashir, F., International Journal of Advanced Manufacturing Technology, 90[5-8] 2017, 1857-1869. https://doi.org/10.1007/s00170-016-9526-z

[228] Nakatani, A., Kitagawa, H., Shimokawa, T., Transactions of the Japan Society of Mechanical Engineers A, 66[643] 2000, 435-441. https://doi.org/10.1299/kikaia.66.435

[229] Shimokawa, T., Nakatani, A., Kitagawa, H., JSME International Journal A, 47[2] 2004, 83-91. https://doi.org/10.1299/jsmea.47.83

[230] Brandl, C., Derlet, P.M., Van Swygenhoven, H., Modelling and Simulation in Materials Science and Engineering, 19[7] 2011, 074005. https://doi.org/10.1088/0965-0393/19/7/074005

[231] Kadau, K., Germann, T.C., Lomdahl, P.S., Holian, B.L., Kadau, D., Entel, P., Kreth, M., Westerhoff, F., Wolf, D.E., Metallurgical and Materials Transactions A, 35[9] 2004, 2719-2723. https://doi.org/10.1007/s11661-004-0217-2

[232] Perron, A., Politano, O., Vignal, V., Philosophical Magazine, 87[1] 2007, 129-145. https://doi.org/10.1080/14786430600936447

[233] An, M., Song, H., Science China: Physics, Mechanics and Astronomy, 56[10] 2013, 1938-1944. https://doi.org/10.1007/s11433-013-5228-9

[234] Mahata, A., Zaeem, M.A., Computational Materials Science, 163, 2019, 176-185. https://doi.org/10.1016/j.commatsci.2019.03.034

[235] Maung, K., Earthman, J.C., Mohamed, F.A., Acta Materialia, 60[16] 2012, 5850-5857. https://doi.org/10.1016/j.actamat.2012.07.026

[236] Li, J.L., Li, S.S., Fan, Z.Z., Li, W., Xiong, Y.C., Chinese Journal of Nonferrous

Metals, 23[5] 2013, 1182-1188. https://doi.org/10.1016/S1003-6326(13)62665-5

[237] Sob, P.B., Alugongo, A.A., Tengen, T.B.B., Advances in Materials Science and Engineering, 2017, 2017, 5418769. https://doi.org/10.1155/2017/5418769

[238] Xu, W., Dávila, L.P., Materials Science and Engineering A, 710, 2018, 413-418. https://doi.org/10.1016/j.msea.2017.10.021

[239] Liu, L., Hou, Z., Tian, Z., Wang, Z., Wang, F., Zhao, X., Liu, R., Computational Materials Science, 156, 2019, 1-6. https://doi.org/10.1016/j.commatsci.2018.09.036

[240] Wang, T.H., Fang, T.H., Kang, S.H., Current Nanoscience, 6[2] 2010, 173-177. https://doi.org/10.2174/157341310790945641

[241] Ma, B., Shi, K., Shang, H., Qi, J., Li, R., Li, G., Journal of Nano Research, 48, 2017, 204-210. https://doi.org/10.4028/www.scientific.net/JNanoR.48.204

[242] Basariya, M.R., Roy, R.K., Pramanick, A.K., Srivastava, V.C., Mukhopadhyay, N.K., Materials Science and Engineering A, 638, 2015, 282-288. https://doi.org/10.1016/j.msea.2015.04.076

[243] Basariya, M.R., Srivastava, V.C., Mukhopadhyay, N.K., Philosophical Magazine, 96[23] 2016, 2445-2456. https://doi.org/10.1080/14786435.2016.1204474

[244] Basariya, M.R., Mukhopadhyay, N.K., Sripathi, S., Padmanabhan, K.A., Journal of Alloys and Compounds, 673, 2016, 199-204. https://doi.org/10.1016/j.jallcom.2016.02.258

[245] Huller, M., Vlcek, J., Dinkel, M., Hoppel, H.W., Goken, M., Philosophical Magazine, 88[8] 2008, 1209-1226. https://doi.org/10.1080/14786430802089854

[246] Okazaki, K., Materials Science and Engineering A, 287[2] 2000, 189-197. https://doi.org/10.1016/S0921-5093(00)00775-9

[247] Nayak, S.S., Wollgarten, M., Banhart, J., Pabi, S.K., Murty, B.S., Materials Science and Engineering A, 527[9] 2010, 2370-2378. https://doi.org/10.1016/j.msea.2009.12.044

[248] Shanmugasundaram, T., Heilmaier, M., Murty, B.S., Sarma, V.S., Materials Science and Engineering A, 527[29-30] 2010, 7821-7825. https://doi.org/10.1016/j.msea.2010.08.070

[249] Sohn, Y.H., Patterson, T., Hofmeister, C., Kammerer, C., Mohr, W., Van Den Bergh, M., Shaeffer, M., Seaman, J., Cho, K., JOM, 64[2] 2012, 234-238. https://doi.org/10.1007/s11837-012-0249-9

[250] Topping, T.D., Ahn, B., Li, Y., Nutt, S.R., Lavernia, E.J., Metallurgical and

Materials Transactions A, 43[2] 2012, 505-519. https://doi.org/10.1007/s11661-011-0849-y

[251] Mukhopadhyay, N.K., Ali, F., Scudino, S., Samadi Khoshkhoo, M., Stoica, M., Srivastava, V.C., Uhlenwinkel, V., Vaughan, G., Suryanarayana, C., Eckert, J., Applied Physics Letters, 103[20] 2013, 201914. https://doi.org/10.1063/1.4831737

[252] Huang, Y., IOP Conference Series - Materials Science and Engineering, 219[1] 2017, 012029. https://doi.org/10.1088/1757-899X/219/1/012029

[253] Schiøtz, J., Vegge, T., Di Tolla, F.D., Jacobsen, K.W., Physical Review B, 60[17] 1999, 11971-11983. https://doi.org/10.1103/PhysRevB.60.11971

[254] Schiøtz, J., Di Tolla, F.D., Jacobsen, K.W., Nature, 391[6667] 1998, 561-563. https://doi.org/10.1038/35328

[255] Zhang, F., Liu, Z., Zhou, J., Materials Letters, 183, 2016, 261-264. https://doi.org/10.1016/j.matlet.2016.07.122

[256] Zhang, Z., Chen, P., Qin, F., Proceedings of the 19th International Conference on Electronic Packaging Technology, ICEPT, 8480604, 2018, 228-232.

[257] Sun, X., Reglero, R., Sun, X., Yacaman, M.J., Materials Chemistry and Physics, 63[1] 2000, 82-87. https://doi.org/10.1016/S0254-0584(99)00214-X

[258] Narayan, J., Journal of Nanoparticle Research, 2[1] 2000, 91-96. https://doi.org/10.1023/A:1010008827415

[259] Liu, Y.W., Zhou, R., Lü, S.Q., International Journal of Nonlinear Sciences and Numerical Simulation, 3[3-4] 2002, 531-534. https://doi.org/10.1515/IJNSNS.2002.3.3-4.531

[260] Hakamada, M., Nakamoto, Y., Matsumoto, H., Iwasaki, H., Chen, Y., Kusuda, H., Mabuchi, M., Materials Science and Engineering A, 457[1-2] 2007, 120-126. https://doi.org/10.1016/j.msea.2006.12.101

[261] Zhang, H., Jiang, Z., Lian, J., Rare Metal Materials and Engineering, 37[2] 2008, 346-349. https://doi.org/10.1016/S1875-5372(10)60008-2

[262] Mohammadabadi, A.S., Dehghani, K., Journal of Materials Engineering and Performance, 17[5] 2008, 662-666. https://doi.org/10.1007/s11665-008-9206-8

[263] Spaepen, F., Yu, D.Y.W., Scripta Materialia, 50[6] 2004, 729-732. https://doi.org/10.1016/j.scriptamat.2003.11.038

[264] Scattergood, R.O., Koch, C.C., Scripta Metallurgica et Materialia, 27[9] 1992, 1195-1200. https://doi.org/10.1016/0956-716X(92)90598-9

[265] Schiøtz, J., Scripta Materialia, 51[8S] 2004, 837-841.
https://doi.org/10.1016/j.scriptamat.2004.05.013

[266] Capolungo, L., Jochum, C., Cherkaoui, M., Qu, J., International Journal of
Plasticity, 21[1] 2005, 67-82. https://doi.org/10.1016/j.ijplas.2004.02.002

[267] Mercier, S., Molinari, A., Estrin, Y., Journal of Materials Science, 42[5] 2007,
1455-1465. https://doi.org/10.1007/s10853-006-0670-y

[268] Capolungo, L., Cherkaoui, M., Qu, J., International Journal of Plasticity, 23[4]
2007, 561-591. https://doi.org/10.1016/j.ijplas.2006.05.003

[269] Benkassem, S., Capolungo, L., Cherkaoui, M., Acta Materialia, 55[10] 2007, .
3563-3572. https://doi.org/10.1016/j.actamat.2007.02.010

[270] Jérusalem, A., Stainier, L., Radovitzky, R., Philosophical Magazine, 87[17] 2007,
2541-2559. https://doi.org/10.1080/14786430701230252

[271] Lan, J., Hong, Y., Archive of Applied Mechanics, 78[6] 2008, 465-476.
https://doi.org/10.1007/s00419-007-0168-3

[272] Barai, P., Weng, G.J., International Journal of Plasticity, 25[12] 2009, 2410-2434.
https://doi.org/10.1016/j.ijplas.2009.04.001

[273] Aifantis, K.E., Konstantinidis, A.A., Materials Science and Engineering A, 503[1-
2] 2009, 198-201. https://doi.org/10.1016/j.msea.2008.04.084

[274] Vo, N.Q., Averback, R.S., Bellon, P., Caro, A., Scripta Materialia, 61[1] 2009, 76-
79. https://doi.org/10.1016/j.scriptamat.2009.03.003

[275] Dongare, A.M., Rajendran, A.M., Lamattina, B., Brenner, D.W., Zikry, M.A.,
Metallurgical and Materials Transactions A, 41[2] 2010, 523-531.
https://doi.org/10.1007/s11661-009-0113-x

[276] Zhang, J.Y., Liu, G., Wang, R.H., Li, J., Sun, J., Ma, E., Physical Review B,
81[17] 2010, 172104. https://doi.org/10.1103/PhysRevB.81.172104

[277] Dongare, A.M., Rajendran, A.M., Lamattina, B., Zikry, M.A., Brenner, D.W.,
Computers, Materials and Continua, 24[1] 2011, 43-60.

[278] Choi, Y., Park, Y., Hyun, S., Physics Letters A, 376[5] 2012, 758-762.
https://doi.org/10.1016/j.physleta.2011.12.027

[279] Nedoushan, R.J., Farzin, M., Mashayekhi, M., Journal of Nano Research, 17,
2012, 35-51. https://doi.org/10.4028/www.scientific.net/JNanoR.17.35

[280] Marchenko, A., Zhang, H., Metallurgical and Materials Transactions A, 43[10]

2012, 3547-3555. https://doi.org/10.1007/s11661-012-1208-3

[281] Pei, L., Lu, C., Tieu, K., Zhu, H., Zhao, X., Cheng, K., Zhang, L., Journal of Nano Research, 25, 2013, 188-194.

[282] Pei, L., Lu, C., Tieu, K., Zhu, H., Zhao, X., Cheng, K., Zhang, L., Journal of Nano Research, 23, 2013, 50-56. https://doi.org/10.4028/www.scientific.net/JNanoR.23.50

[283] Galanis, N.V., Remediakis, I.N., Kopidakis, G., Mechanics of Materials, 67, 2013, 79-85. https://doi.org/10.1016/j.mechmat.2013.07.019

[284] Wen, P., Tao, G., Ren, B.X., Pei, Z., Acta Physica Sinica, 64[12] 2015, 126201.

[285] Huang, C.C., Chiang, T.C., Fang, T.H., Applied Surface Science, 353, 2015, 494-498. https://doi.org/10.1016/j.apsusc.2015.06.135

[286] Yuan, R., Beyerlein, I.J., Zhou, C., Scientific Reports, 6, 2016, 26254. https://doi.org/10.1038/srep26254

[287] Fang, T.H., Huang, C.C., Chiang, T.C., Materials Science and Engineering A, 671, 2016, 1-6. https://doi.org/10.1016/j.msea.2016.06.042

[288] Rida, A., Rouhaud, E., Makke, A., Micoulaut, M., Mantisi, B., Philosophical Magazine, 97[27] 2017, 2387-2405. https://doi.org/10.1080/14786435.2017.1334136

[289] Chandross, M., Curry, J.F., Babuska, T.F., Lu, P., Furnish, T.A., Kustas, A.B., Nation, B.L., Staats, W.L., Argibay, N., Scripta Materialia, 143, 2018, 54-58. https://doi.org/10.1016/j.scriptamat.2017.09.006

[290] Meyers, M.A., Chawla, K.K., Mechanical Behavior of Materials, Cambridge University Press, 2009. https://doi.org/10.1017/CBO9780511810947

[291] Ray, K.K., Chakraborty, K., Journal of Materials Science Letters, 15[8] 1996, 727-730. https://doi.org/10.1007/BF00264128

[292] Phillips, W.L., Armstrong, R.W., Metallurgical Transactions, 3[10] 1972, 2571-2577. https://doi.org/10.1007/BF02644231

[293] Nakao, Y., Miura, H., Sakai, T., Materials Scince Forum, 558-559, 2007, 1329-1334. https://doi.org/10.4028/www.scientific.net/MSF.558-559.1329

[294] Pasebani, S., Toroghinejad, M.R., Materials Science and Engineering A, 527[3] 2010, 491-497. https://doi.org/10.1016/j.msea.2009.09.029

[295] Zhao, Y.H., Liao, X.Z., Horita, Z., Langdon, T.G., Zhu, Y.T., Materials Science and Engineering A, 493[1-2] 2008, 123-129. https://doi.org/10.1016/j.msea.2007.11.074

[296] Balogh, L.U.T., Zhao, Y., Zhu, Y.T., Horita, Z., Xu, C., Langdon, T.G., Acta Materialia, 56[4] 2008, 809-820. https://doi.org/10.1016/j.actamat.2007.10.053

[297] Sun, H., Shi, Y., Journal of Materials Science and Technology, 25[3] 2009, 347-350.

[298] Bahmanpour, H., Youssef, K.M., Horky, J., Setman, D., Atwater, M.A., Zehetbauer, M.J., Scattergood, R.O., Koch, C.C., Acta Materialia, 60[8] 2012, 3340-3349. https://doi.org/10.1016/j.actamat.2012.02.036

[299] Wang, N., Wang, Z., Aust, K.T., Erb, U., Materials Science and Engineering A, 237[2] 1997, 150-158. https://doi.org/10.1016/S0921-5093(97)00124-X

[300] Ebrahimi, E., Ahmed, Z., Morgan, K.L., Materials Research Society Symposium-Proceedings, 2001, 634.

[301] Jeong, D.H., Gonzalez, F., Palumbo, G., Aust, K.T., Erb, U., Scripta Materialia, 44[3] 2001, 493-499. https://doi.org/10.1016/S1359-6462(00)00625-4

[302] Schuh, C.A., Nieh, T.G., Yamasaki, T., Scripta Materialia, 46[10] 2002, 735-740. https://doi.org/10.1016/S1359-6462(02)00062-3

[303] Ebrahimi, F., Bourne, G.R., Kelly, M.S., Matthews, T.E., Nanostructed Materials, 11[3] 1999, 343-350. https://doi.org/10.1016/S0965-9773(99)00050-1

[304] Hughes, G.D., Smith, S.D., Pande, C.S., Johnson, H.R., Armstrong, R.W., Scripta Metallurgica, 20[1] 1986, 93-97. https://doi.org/10.1016/0036-9748(86)90219-X

[305] Erb, U.. Nanostructured Materials, 6[5-8] 1995, 533-538. https://doi.org/10.1016/0965-9773(95)00114-X

[306] El-Sharik, A.M., Erb, U., Palumbo, G., Aust, K.T., Scripta Metallurgica et Materialia, 27[9] 1992, 1185-1188. https://doi.org/10.1016/0956-716X(92)90596-7

[307] Zhao, Y., Shen, T.D., Zhang, J., Nanoscale Research Letters, 2[10] 2007, 476-491. https://doi.org/10.1007/s11671-007-9095-z

[308] Koslowski, M., Lee, D.W., Lei, L., Journal of the Mechanics and Physics of Solids, 59[7] 2011, 1427-1436. https://doi.org/10.1016/j.jmps.2011.03.011

[309] Cao, L., Koslowski, M., Acta Materialia, 61[4] 2013, 1413-1420. https://doi.org/10.1016/j.actamat.2012.11.018

[310] Liu, X., Yuan, F., Wei, Y., Applied Surface Science, 279, 2013, 159-166. https://doi.org/10.1016/j.apsusc.2013.04.062

[311] Chamani, M., Farrahi, G.H., Movahhedy, M.R., Tribology International, 107,

2017, 18-24. https://doi.org/10.1016/j.triboint.2016.11.020

[312] Liu, H., Zhou, J., Materials Letters, 163, 2016, 179-182.
https://doi.org/10.1016/j.matlet.2015.10.068

[313] Cao, Z.H., Wang, L., Hu, K., Huang, Y.L., Meng, X.K., Acta Materialia, 60[19]
2012, 6742-6754. https://doi.org/10.1016/j.actamat.2012.08.047

[314] Jeng, Y.R., Tsai, P.C., Chiang, S.H., Wear, 303[1-2] 2013, 262-268.
https://doi.org/10.1016/j.wear.2013.02.019

[315] Volpp, T., Göring, E., Kuschke, W.M., Arzt, E., Nanostructured Materials, 8[7]
1997, 855-865. https://doi.org/10.1016/S0965-9773(98)00019-1

[316] Oh-ishi, K., Horita, Z., Smith, D.J., Valiev, R.Z., Nemoto, M., Langdon, T.G.,
Journal of Materials Research, 14[11] 1999, 4200-4207.
https://doi.org/10.1557/JMR.1999.0569

[317] Abdel-Karim, R., Reda, Y., Muhammed, M., El-Raghy, S., Shoeib, M., Ahmed,
H., Journal of Nanomaterials, 2011, 2011, 519274.
https://doi.org/10.1155/2011/519274

[318] Cheung, C., Djuanda, F., Erb, U., Palumbo, G., Nanostructured Materials, 5[5]
1995, 513-523. https://doi.org/10.1016/0965-9773(95)00264-F

[319] Giallonardo, J.D., Erb, U., Aust, K.T., Palumbo, G., Philosophical Magazine,
91[36] 2011, 4594-4605. https://doi.org/10.1080/14786435.2011.615350

[320] Vicenzo, A., Journal of the Electrochemical Society, 160[11] 2013, D570-D577.
https://doi.org/10.1149/2.109311jes

[321] Desai, J.A., Kumar, A., Metals and Materials International, 22[3] 2016, 451-458.
https://doi.org/10.1007/s12540-016-5644-2

[322] Hoffmann, J.E., Schmitt, M.T., Eifler, D., Beck, T., Klär, P., Saumer, M.,
Materials Testing, 60[11] 2018, 1041-1049. https://doi.org/10.3139/120.111259

[323] Matsui, I., Kanetake, M., Mori, H., Takigawa, Y., Higashi, K., Materials Letters,
205, 2017, 211-214. https://doi.org/10.1016/j.matlet.2017.06.094

[324] Sriraman, K.R., Raman, S.G.S., Seshadri, S.K., Materials Science and
Technology, 22[1] 2006, 14-20. https://doi.org/10.1179/174328406X79243

[325] Chassaing, E., Portail, N., Levy, A.F., Wang, G., Journal of Applied
Electrochemistry, 34[11], 2004, 1085-1091. https://doi.org/10.1007/s10800-004-
2460-z

[326] Bigos, A., Beltowska-Lehman, E., Kot, M., Surface and Coatings Technology, 317, 2017, 103-109. https://doi.org/10.1016/j.surfcoat.2017.03.036

[327] Abrosimova, G.E., Aronin, A.S., Zverkova, I.I., Gurov, A.F., Kiryanov, Y.V., Physics of the Solid State, 40[1] 1998, 8-13. https://doi.org/10.1134/1.1130536

[328] Prasad, M.J.N.V., Chokshi, A.H., Kovove Materialy, 49[1] 2011, 93-99.

[329] Shi, J., Cao, Z.H., Hu, K., Pan, G.J., Wei, M.Z., Xu, L.J., Meng, X.K., Materials Letters, 115, 2014, 79-81. https://doi.org/10.1016/j.matlet.2013.10.024

[330] Schuh, C.A., Nieh, T.G., Iwasaki, H., Acta Materialia, 51[2] 2003, 431-443. https://doi.org/10.1016/S1359-6454(02)00427-5

[331] Giga, A., Kimoto, Y., Takigawa, Y., Higashi, K., Scripta Materialia, 55[2] 2006, 143-146. https://doi.org/10.1016/j.scriptamat.2006.03.047

[332] Sriraman, K.R., Raman, S.G.S., Seshadri, S.K., Materials Science and Engineering A, 418[1-2] 2006, 303-311. https://doi.org/10.1016/j.msea.2005.11.046

[333] Sriraman, K.R., Raman, S.G.S., Seshadri, S.K., Materials Letters, 61[3] 2007, 715-718. https://doi.org/10.1016/j.matlet.2006.05.049

[334] Detor, A.J., Schuh, C.A., Acta Materialia, 55[1] 2007, 371-379. https://doi.org/10.1016/j.actamat.2006.08.032

[335] Trelewicz, J.R., Schuh, C.A., Acta Materialia, 55[17] 2007, 5948-5958. https://doi.org/10.1016/j.actamat.2007.07.020

[336] Trelewicz, J.R., Schuh, C.A., Applied Physics Letters, 93[17] 2008, 171916. https://doi.org/10.1063/1.3000655

[337] Trelewicz, J.R., Schuh, C.A., Scripta Materialia, 61[11] 2009, 1056-1059. https://doi.org/10.1016/j.scriptamat.2009.08.026

[338] Jang, D., Greer, J.R., Scripta Materialia, 64[1] 2011, 77-80. https://doi.org/10.1016/j.scriptamat.2010.09.010

[339] Rupert, T.J., Trelewicz, J.R., Schuh, C.A., Journal of Materials Research, 27[9] 2012, 1285-1294. https://doi.org/10.1557/jmr.2012.55

[340] Nia, N.S., Savall, C., Creus, J., Bourgon, J., Girault, P., Metsue, A., Cohendoz, S., Feaugas, X., Materials Science and Engineering A, 678, 2016, 204-214. https://doi.org/10.1016/j.msea.2016.09.097

[341] Ong, C.Y.A., Blackwood, D.J., Li, Y., Surface and Coatings Technology, 357, 2019, 23-27. https://doi.org/10.1016/j.surfcoat.2018.09.086

[342] Nagai, T., Hodouchi, K., Matsubara, H., Surface and Coatings Technology, 253, 2014, 109-114. https://doi.org/10.1016/j.surfcoat.2014.05.022

[343] Liu, X., Yuan, F., Lu, K., Wei, W., Journal of Materials Science and Technology, 12[6] 1996, 409-412.

[344] Khaliq, M.W., Butt, M.Z., Saleem, M., Materials Research Express, 4[7] 2017, 076513. https://doi.org/10.1088/2053-1591/aa7b3a

[345] Gobien, J.M., Scattergood, R.O., Goodwin, F.E., Koch, C.C., Materials Science and Engineering A, 518[1-2] 2009, 84-88. https://doi.org/10.1016/j.msea.2009.04.023

[346] Narayan, J., Venkatesan, R.K., Kvit, A., Journal of Nanoparticle Research, 4[3] 2002, 265-269. https://doi.org/10.1023/A:1019925315398

[347] Conrad, H., Narayan, J., Applied Physics Letters, 81[12] 2002, 2241-2243. https://doi.org/10.1063/1.1507353

[348] Choi, H.J., Kim, Y., Shin, J.H., Bae, D.H., Materials Science and Engineering A, 527[6] 2010, 1565-1570. https://doi.org/10.1016/j.msea.2009.10.035

[349] Wu, G., Chan, K.C., Zhu, L., Sun, L., Lu, J., Nature, 545[7652] 2017, 80-83. https://doi.org/10.1038/nature21691

[350] Lu, J., Jin, L., Dong, J., Zeng, X.Q., Ding, W.J., Yao, Z.Y., Chinese Journal of Nonferrous Metals, 19[3] 2009, 424-432.

[351] Song, H.Y., An, M.R., Li, Y.L., Deng, Q., Journal of Applied Physics, 116[21] 2014, 214305. https://doi.org/10.1063/1.4903526

[352] Wang, K.Y., Shen, T.D., Quan, M.X., Wei, W.D., Journal of Materials Science Letters, 12[23] 1993, 1818-1820. https://doi.org/10.1007/BF00539997

[353] Pachla, W., Kulczyk, M., Sus-Ryszkowska, M., Mazur, A., Kurzydlowski, K.J., Journal of Materials Processing Technology, 205[1-3] 2008, 173-182. https://doi.org/10.1016/j.jmatprotec.2007.11.103

[354] Chang, L., Zhou, C.Y., Li, J., He, X.H., Computational Materials Science, 142, 2018, 135-144. https://doi.org/10.1016/j.commatsci.2017.10.017

[355] Senkov, O.N., Srisukhumbowornchai, N., Öveçoglu, M.L., Froes, F.H., Journal of Materials Research, 13[12] 1998, 3399-3410. https://doi.org/10.1557/JMR.1998.0463

[356] Glezer, A.M., Shurygina, N.A., Blinova, E.N., Permyakova, I.E., Firstov, S.A., Journal of Materials Science and Technology, 31[1] 2015, 91-96. https://doi.org/10.1016/j.jmst.2014.09.006

[357] Glezer, A.M., Shurygina, N.A., Blinova, E.N., Permyakova, I.E., Firstov, S.A., Russian Metallurgy, 2012[10] 2012, 853-859. https://doi.org/10.1134/S0036029512100060

[358] Wang, B., Tomar, V., Haque, A., Materials Letters, 152, 2015, 105-108. https://doi.org/10.1016/j.matlet.2015.03.105

[359] Ruestes, C.J., Bertolino, G., Ruda, M., Farkas, D., Bringa, E.M., Scripta Materialia, 71, 2014, 9-12. https://doi.org/10.1016/j.scriptamat.2013.09.010

[360] Alves, H., Ferreira, M., Köster, U., Müller, B., Materials Science Forum, 225-227[2] 1996, 769-774. https://doi.org/10.4028/www.scientific.net/MSF.225-227.769

[361] Borroto, A., Bruyère, S., Thurieau, N., Gendarme, C., Jimenez-Piqué, E., Roa, J.J., Pierson, J.F., Mücklich, F., Horwat, D., Journal of Alloys and Compounds, 729, 2017, 137-143. https://doi.org/10.1016/j.jallcom.2017.09.153

[362] Zheng, G.P., Acta Materialia, 55[1] 2007, 149-159. https://doi.org/10.1016/j.actamat.2006.07.034

[363] Shi, J., Li, J., Sun, X., Rare Metal Materials and Engineering, 36[5] 2007, 937-940.

[364] Rock, C., Qiu, J., Okazaki, K., Journal of Materials Science, 33[1] 1998, 241-246. https://doi.org/10.1023/A:1004386822343

[365] Meyers, M.A., Benson, D.J., Vöhringer, O., Kad, B.K., Xue, Q., Fu, H.H., Materials Science and Engineering A, 322[1-2] 2002, 194-216. https://doi.org/10.1016/S0921-5093(01)01131-5

[366] Tang, Y., Bringa, E.M., Meyers, M.A., Materials Science and Engineering A, 580, 2013, 414-426. https://doi.org/10.1016/j.msea.2013.05.024

[367] Huang, C., Peng, X., Zhao, Y., Weng, S., Yang, B., Fu, T., Materials Science and Engineering A, 738, 2018, 1-9. https://doi.org/10.1016/j.msea.2018.09.053

[368] Tramontina, D.R., Hahn, E.N., Meyers, M.A., Bringa, E.M., AIP Conference Proceedings, 1793, 2017, 070002.

[369] Huang, C., Peng, X., Fu, T., Chen, X., Xiang, H., Li, Q., Hu, N., Materials Science and Engineering A, 700, 2017, 609-616. https://doi.org/10.1016/j.msea.2017.06.048

[370] Saha, S., Motalab, M.A., Mahboob, M., Computational Materials Science, 136, 2017, 52-59. https://doi.org/10.1016/j.commatsci.2017.04.025

[371] Kimura, Y., Hidaka, H., Takaki, S., Materials Transactions, JIM, 40[10] 1999, 1149-1157. https://doi.org/10.2320/matertrans1989.40.1149

[372] Jang, D., Atzmon, M., Journal of Applied Physics, 93[11] 2003, 9282-9286. https://doi.org/10.1063/1.1569035

[373] Lu, K., Sui, M.L., Scripta Metallurgica et Materiala, 28[12] 1993, 1465-1470. https://doi.org/10.1016/0956-716X(93)90576-E

[374] Jeon, J.B., Lee, B.J., Chang, Y.W., Scripta Materialia, 64[6] 2011, 494-497. https://doi.org/10.1016/j.scriptamat.2010.11.019

[375] Yuan, F.P., Science China: Physics, Mechanics and Astronomy, 55[9] 2012, 1657-1663. https://doi.org/10.1007/s11433-012-4830-6

[376] Wang, P., Xu, J.G., Zhang, Y.G., Song, H.Y., Acta Physica Sinica, 65[23] 2016, 236201.

[377] Kuhr, B.R., Aifantis, K.E., Materials Science and Engineering A, 745, 2019, 107-114. https://doi.org/10.1016/j.msea.2018.12.053

[378] Xu, T., He, S., Journal of Aeronautical Materials, 37[1] 2017, 73-79.

[379] Fang, Q., Li, L., Li, J., Wu, H., Computational Materials Science, 152, 2018, 236-242. https://doi.org/10.1016/j.commatsci.2018.06.001

[380] Tejedor, R., Edalati, K., Benito, J.A., Horita, Z., Cabrera, J.M., Materials Science and Engineering A, 743, 2019, 597-605. https://doi.org/10.1016/j.msea.2018.11.127

[381] Muñoz-Morris, M.A., Garcia Oca, C., Morris, D.G., Acta Materialia, 51[17] 2003, 5187-5197. https://doi.org/10.1016/S1359-6454(03)00382-3

[382] Mhadhbi, M., Khitouni, M., Escoda, L., Suñol, J.J., Dammak, M., Journal of Alloys and Compounds, 509[7] 2011, 3293-3298. https://doi.org/10.1016/j.jallcom.2010.10.214

[383] Li, C.L., La, P.Q., Liu, H., Wei, Y.P., Lu, X.F., Chen, J.M., Transactions of Materials and Heat Treatment, 33[4] 2012, 17-21.

[384] Shen, T.D., Schwarz, R.B., Feng, S., Swadener, J.G., Huang, J.Y., Tang, M., Zhang, J., Vogel, S.C., Zhao, Y., Acta Materialia, 55[15] 2007, 5007-5013. https://doi.org/10.1016/j.actamat.2007.05.018

[385] Li, Y.F., La, P.Q., Wei, Y.P., Lu, Y., Yang, Y., Journal of Iron and Steel Research International, 18[3] 2011, 65-71. https://doi.org/10.1016/S1006-706X(11)60039-3

[386] Ding, B.Z., Tong, H.Y., Jiang, H.G., Wang, J.T., Wei, W.D., Scripta Metallurgica et Materialia, 28[9] 1993, 1107-1112. https://doi.org/10.1016/0956-716X(93)90018-N

[387] Li, J.M., Quan, M.X., Hu, Z.Q., Applied Physics Letters, 69[11] 1996, 1559-1561.

https://doi.org/10.1063/1.117061

[388] Okayasu, M., Ishida, D., Metallurgical and Materials Transactions A, 50[3] 2019, 1380-1388. https://doi.org/10.1007/s11661-018-5083-4

[389] Kashyap, B.P., Tangri, K., Acta Materialia, 45[6] 1997, 2383-2395. https://doi.org/10.1016/S1359-6454(96)00341-2

[390] Hasegawa, M., Yoon, S., Guillonneau, G., Zhang, Y., Frantz, C., Niederberger, C., Weidenkaff, A., Michler, J., Philippe, L., Physical Chemistry Chemical Physics, 16[47] 2014, 26375-26384. https://doi.org/10.1039/C4CP03744H

[391] Chen, X.H., Lu, J., Lu, L., Lu, K., Scripta Materialia, 52[10] 2005, 1039-1044. https://doi.org/10.1016/j.scriptamat.2005.01.023

[392] Liu, X.D., Hu, Z.Q., Ding, B.Z., Nanostructured Materials, 2[5] 1993, 545-552. https://doi.org/10.1016/0965-9773(93)90172-8

[393] Li, S., Lu, K., Guo, F., Chu, R., Wang, Z., Materials Letters, 30[4] 1997, 305-310. https://doi.org/10.1016/S0167-577X(96)00208-X

[394] Glezer, A.M., Manaenkov, S.E., Permyakova, I.E., Shurygina, N.A., Russian Metallurgy, 2011[10] 2011, 947-955. https://doi.org/10.1134/S0036029511100041

[395] Curry, J.F., Babuska, T.F., Furnish, T.A., Lu, P., Adams, D.P., Kustas, A.B., Nation, B.L., Dugger, M.T., Chandross, M., Clark, B.G., Boyce, B.L., Schuh, C.A., Argibay, N., Advanced Materials, 30, 2018, 1802026. https://doi.org/10.1002/adma.201802026

[396] Heckman, N.M., Foiles, S.M., O'Brien, C.J., Chandross, M., Barr, C.M., Argibay, N., Hattar, K., Lu, P., Adams, D.P., Boyce, B.L., Nanoscale, 10[45] 2018, 21231-21243. https://doi.org/10.1039/C8NR06419A

Keyword Index

www.ingramcontent.com/pod-product-compliance
Lightning Source LLC
Chambersburg PA
CBHW071653210326
41597CB00017B/2193